Marvels of Mammals

Unlocking Nature's Secrets

The Science of Biomimetics

Innovations Shaping a Better World

R. K. Little

Table of Contents

Introduction

Considering Earth's biodiversity, mammals stand out as true marvels that embody a spectrum of abilities that span the sublime to the extraordinary. From the silent elegance of a prowling feline and the tenacity of a sprinting cheetah to the intellectual prowess of the mighty human mind, mammals exemplify the pinnacle of nature's design. Yet, beyond the captivating spectacle of their existence lies a deeper narrative—one that unveils a profound secret, a secret that has fueled the imagination of scientists, engineers, and innovators throughout history.

Welcome to the enthralling journey through the pages of *Marvels of Mammals: Unlocking Nature's Secrets*. As you embark on this expedition, prepare to be captivated by the extraordinary tales of mammals, their astonishing abilities, and the ingenious ways in which human minds have harnessed the power of biomimicry to shape a better world.

Imagine a world where every challenge is an opportunity and every obstacle is a mere stepping stone. Nature's infinite wisdom has endowed mammals with extraordinary gifts. From the icy landscapes of the Arctic, where the cunning Arctic fox dons a winter coat as if tailored by nature's own fashion designer, to the steamy jungles where the chameleon exhibits a color palette unmatched by any artist, adaptation is the heartbeat of mammalian survival.

As we journey to unlock these secrets of nature, we will

- delve into the captivating narratives of nature's finest choreographers and witness how the artistry of adaptation has inspired some of the most groundbreaking innovations in human technology. From temperature-regulating fabrics to adaptive camouflage in military applications, the dance of adaptation echoes through the corridors of progress.

- unravel the tales of motion maestros and the tangible echoes of their prowess in the wheels, wings, and propellers that propel us forward. Mammals have mastered the poetry of motion, and in doing so, have become architects of inspiration for the world of transportation.

- experience the symphony of senses that orchestrates a mesmerizing melody within the realm of mammals. But the story does not end with this magnificent opus; it evolves into a new crescendo as human innovation seeks to replicate and enhance our sensory capacities.

- reveal a profound understanding of community dynamics uncovered in the intricate web of social structures woven by mammals—be it the familial bonds of elephants, the cooperative societies of meerkats, or the complex hierarchies of primates.

- find inspiration in the intricate social architectures of our mammalian counterparts. From organizational structures modeled after wolf packs to the development of collaborative technologies inspired by the wisdom of ants, biomimicry breathes life into our communal aspirations.

- uncover the unrivaled cognitive prowess of certain species—most notably, humans. From the problem-solving prowess of the octopus to the strategic intellect of the chimpanzee, the cognitive odyssey of mammals unfolds in a tapestry of ingenious minds.

- discover the inspiration drawn from nature's cognitive architects. From artificial intelligence algorithms inspired by the learning patterns of dolphins to the development of robotic prosthetics guided by the elegance of neural connections, mammalian minds continue to shape the frontiers of human innovation.

- illuminate the phenomenon of symbiosis, where diverse organisms forge alliances for mutual benefit—a testament to the alchemy of collaboration in nature. The grand tapestry of life is woven not just with the threads of competition but also with the delicate strands of cooperation.

- dive into the heart of the action where nature's genius isn't just a showstopper in the wild but takes center stage in shaping how we live and heal.

- learn how mammals' resilient defenses inspire game-changing treatments, precision medicine, and smart ways to prevent diseases. We're talking about breakthroughs in medicine that make us rethink how we combat diseases.

- explore the bigger picture, zooming out from the micro-world of medicine to the wild dance of life. Nature's got this amazing rhythm, a harmony where different species rely on each other to keep the balance.

- give a big cheer for these awesome partnerships and celebrate the idea that when different life forms coexist and work together, incredible things happen.

Join us in this celebration of nature's genius. Imagine a world where the lessons from the wild aren't only fascinating stories but are also blueprints for a better future. *Marvels of Mammals* invites you to be a part of this journey where the incredible abilities of mammals are wonders to admire and guides to steer us toward a more harmonious life on this planet. It's a ride that connects the untamed with the everyday, where nature becomes a bridge to a future where humans and the wild live together in harmony. Let's ride this wave of discovery and make it our own.

Chapter 1:

The Super Sniffers—A Dog's Sense of Smell

Enter the olfactory wonderland of dogs, where their sense of smell surpasses human comprehension. A dog's nose can do more than you think. Step into a realm where scent is not just a fragrance but a symphony, and the canine nose is an extraordinary maestro. Our furry friends, the Super Sniffers, have olfactory prowess that is a marvel in itself. From deciphering the secrets of scent trails to detecting diseases with unparalleled accuracy, a dog's nose is a beacon of biological brilliance. Let's explore how the fascinating world of biomimetics harnesses these canine capabilities, offering mankind a glimpse into the realm of heightened senses and innovative solutions.

The Anatomy of Olfaction—The Remarkable Olfactory System

In the symphony of sensory experiences that define the animal kingdom, the olfactory prowess of certain mammals stands as a masterpiece. A dog's olfactory system is a marvel of efficiency and complexity. Picture the olfactory bulb as the beating heart of their scent-centric universe. Nestled within their snouts, these olfactory powerhouses vary in size across species, with some exceeding our expectations tenfold. It's not just about size, though; it's about the intricate structure that turns the canine nose into a veritable scent-processing factory. It's a structural dance that allows them to detect scents at levels that leave humans in awe.

Scent Receptors at Work

As we journey deeper into the olfactory wonderland, our attention shifts to the unsung heroes: the scent receptors. Dogs possess a staggering number of these tiny molecular detectors, each tuned to specific scent molecules. This is no ordinary sniffing; it's an intricate interplay of receptors that transform scent into a multisensory experience. It's a saga of sensitivity that goes beyond our understanding, hinting at a world where every breeze, every waft, carries a story waiting to be unraveled.

The Brain's Olfactory Processing and Scent Discrimination Abilities

But how good is an intricate olfactory system without a brain that can interpret the olfactory opera? The canine brain's olfactory processing is nothing short of extraordinary, showcasing scent discrimination abilities that put our olfactory efforts to shame. A dog can remember and distinguish an astonishing array of smells. With their scent-tracking prowess, a dog can not only survive in the wild but also find remarkable applications in human fields, ranging from law enforcement to medical diagnostics.

Applications of Canine Olfaction

Innovations in Electronic Nose Technology

Bridging the gap between nature and innovation, the exploration of canine olfaction has birthed a revolution in electronic nose technology. Some advancements have arisen from studying the biological marvel of a dog's nose. Electronic noses, inspired by the intricate workings of canine olfaction, have transcended mere imitation to become powerful tools in various fields. These artificial noses are not merely mimicking the dogs; they are evolving beyond their biological counterparts. With applications ranging from detecting explosives at airports to diagnosing diseases through breath analysis, electronic noses are reshaping industries.

The Future of Smell-Based Technology

The Super Sniffers lay the groundwork for a future where the boundaries between man and mammal blur in the pursuit of innovation. As we stand at the crossroads of nature's wonders and human innovation, dogs beckon us toward a future where scent-based technologies redefine the boundaries of what's possible.

Many potential breakthroughs could redefine our world. As we stand on the precipice of a new era in biomimetics, where the wonders of mammalian abilities guide our technological advancements, scientists are eager for more.

Medical Detection and Diagnostics

In the realm of healthcare, a revolution is brewing, and it's all about scent. Imagine a world where diseases can be detected not through invasive procedures but through the subtle nuances carried in our breath. From the early detection of diseases like cancer and diabetes to the identification of infectious agents, the olfactory blueprint of these canines becomes a guiding light for the future of medicine. The convergence of biology and technology promises a new era where man's best friend inspires innovations that save lives and redefine healthcare as we know it.

Environmental Monitoring

The planet is sending distress signals, and dogs offer a unique perspective on how we can respond. These great mammals, whose noses are finely attuned to the scents of the Earth, guide us toward a future where technology harmonizes with nature to safeguard the delicate balance of our planet.

Enhancing Food Safety

Our journey into the future wouldn't be complete without a tantalizing exploration of the role scent-based technology plays in ensuring the safety and quality of the food we consume. Remember the electronic noses mentioned before? By mimicking the olfactory precision of dogs, they are becoming instrumental in detecting contaminants and spoilage and even assessing the freshness of produce. A dog's ability to discern the subtlest changes in scent finds application in the agricultural and food industries, promising a future where our meals are not only delicious but also reliably safe. As we peer into the future of smell-based technology, the legacy of dogs becomes a guiding force for innovation.

As we venture into the next realm of marvels, remember that the wonders of the animal kingdom are not confined to a single sense. Dogs have merely opened the door to a menagerie of extraordinary abilities that nature has bestowed upon mammals. The Super Sniffers have paved the way, and now we journey into the uncharted territories of astonishing adaptations and biomimetic breakthroughs that promise to reshape our understanding of the marvelous mammals that share this planet with us.

Chapter 2:

Purrfection Unleashed—

Exploring the Marvels of

Grooming and Whiskers in

Cats

Cats are masters of self-care, spending a significant portion of their waking hours meticulously grooming their fur. But what makes their grooming prowess so extraordinary? It turns out that nature has equipped them with an ingenious tool: their tongue.

Ever wonder why a cat's tongue feels like Velcro? If you've ever witnessed a cat groom itself, you can't help but marvel at the unique texture of its tongue; it's akin to a miniature strip of the famous hook-and-loop fastener. Let's step into the realm of feline grooming, uncovering the secrets hidden within the structure and function of a cat's tongue.

As we delve into the intricacies of this seemingly unassuming organ, we discover a world of biomimetics. Let this purrfectly fascinating journey ignite your curiosity about the incredible abilities found in the animal kingdom—abilities that continue to inspire and shape the technological landscape of our human endeavors. The wonders of grooming are not just confined to the tongue, whiskers, and fur of our feline friends; they extend into the very fabric of human innovation.

The Cat's Unique Tongue Structure

In the feline realm, a cat's tongue is a marvel of nature, a multifunctional tool designed with precision and purpose. It is the intricate details that make it a masterpiece of biological engineering.

Papillae and Their Functions

At the heart of this grooming marvel lies a carpet of tiny wonders known as papillae. These microscopic, hook-like structures cover the surface of a cat's tongue, serving as the ultimate grooming apparatus. Imagine a living Velcro, designed to trap loose fur, debris, and even those pesky parasites that dare to invade a cat's luxurious coat.

You can brush your fingers against the soft fur of your feline friend, but it's the papillae that go beyond the surface, navigating through the strands to keep the coat pristine. This intricate design not only aids in maintaining a cat's immaculate appearance but also plays a vital role in their survival, ensuring a stealthy approach in the wild by eliminating any compromising scents.

Self-Cleaning Mechanisms and Saliva Dispensing

A cat's grooming prowess goes beyond merely collecting unwanted hitchhikers. The tongue itself is a self-cleaning marvel, equipped with a built-in mechanism that helps it shed any accumulated dirt or loose fur. As your cat meticulously licks its fur, the papillae act as self-cleaning combs, untangling knots and removing any foreign particles.

But that's not all. There's also the cat's saliva. Beyond being a simple moisturizer, a cat's saliva carries enzymes with antibacterial properties. As your feline friend indulges in grooming, saliva is dispensed generously, creating a natural protective barrier against infections and promoting the healing of minor wounds.

Inspiring Innovations

The enchanting world of feline grooming doesn't just captivate us with its natural wonders; it sparks a revolution in human innovation. The cat's tongue, a biological masterpiece, becomes a muse for brilliant minds across various fields. Engineers and designers draw inspiration from this feline masterpiece, seeking solutions that not only enhance efficiency but also contribute to sustainability. The cat's tongue becomes not just a grooming tool but a source of inspiration, a reminder that in the animal kingdom, nature holds the keys to innovation.

Industrial Applications

In the realm of industry, the meticulous design of the cat's tongue finds a new purpose. Its papillae, self-cleaning mechanism, and saliva-dispensing abilities serve as a blueprint for cutting-edge biomimetics. Inspired by self-cleaning mechanisms and efficient grooming capabilities, industrial engineers are crafting surfaces that shrug off contaminants effortlessly. Imagine machinery that not only operates with precision but also maintains itself, shedding debris and dirt just like a vigilant feline.

From factories to spacecraft, the cat's tongue biomimetics offers a glimpse into a future where maintenance becomes a seamless, automated process. The relentless pursuit of efficiency takes a cue from nature's engineer—the cat.

Medical Device Design

The cat's tongue becomes a beacon of inspiration. The self-cleaning prowess and antibacterial properties of feline saliva pave the way for cutting-edge medical device designs. Think of catheters and implants that not only perform their intended functions but also resist bacterial colonization, reducing the risk of infections.

As we delve into the crossroads of biology and technology, the cat's tongue guides the hands of medical pioneers, shaping a future where our health is safeguarded by lessons learned from our four-legged companions.

Future Possibilities

The journey into biomimetics is an ever-evolving adventure, and the cat's tongue opens doors to possibilities yet unexplored. Picture a world where robotics seamlessly mimic feline grooming precision, from delicate surgeries to intricate tasks in hazardous environments. The future holds the promise of innovations that blend nature's elegance with human ingenuity.

Beyond the immediate applications, biomimetics inspired by the cat's tongue may lead to breakthroughs in environmental conservation, material science, and beyond. The feline muse becomes a guiding light, encouraging us to envision a future where the marvels of nature not only inspire but actively shape the world we live in.

Whiskers and Whispers—

Biomimetics of a Cat's Whiskers

Why are a cat's whiskers so sensitive? In the intricate tapestry of the animal kingdom, few features captivate our curiosity quite like the mystical marvels of a cat's whiskers. These delicate, seemingly unassuming strands are not mere adornments; they are nature's masterpieces, finely tuned receptors that unlock a world of secrets through touch and vibration. There are secrets encoded in every twitch, a silent language spoken through every delicate touch. Once we unlock the hidden marvels inspired by the sensory brilliance of a cat's whiskers, your understanding of these seemingly simple appendages will never be the same!

Sensory Receptors in Whiskers

At first glance, whiskers may seem like nothing more than an aesthetic flourish on a cat's face, but dive deeper, and you'll uncover a masterpiece of biological engineering. Cat whiskers, scientifically known as vibrissae, are equipped with a network of sensory receptors that put even the most sophisticated human-made sensors to shame.

Each whisker is a finely tuned instrument, housing a treasure trove of nerve endings at its base. These receptors are exquisitely attuned to the slightest changes in the air, temperature, and even the most imperceptible vibrations. It's as if nature gifted our feline friends with an extrasensory perception that we can only marvel at.

Detecting Changes in the Environment

Imagine a world where whispers of the wind carry vital information directly to your nervous system. For cats, this is not a fantasy but a daily reality. Whiskers serve as nature's antennae, effortlessly detecting changes in air currents and alerting them to the approach of prey or potential threats.

In the silent darkness of night, a cat's whiskers become a beacon of perception, guiding them through the shadows with unparalleled precision. The subtle dance of their whiskers grants them an almost supernatural ability to navigate the world, making each step a testament to the marvels of ingenuity.

Communicating Through Whiskers

Whiskers are not just silent observers; they are also the language of feline communication. Watch closely and you'll witness a secret code unfolding in the subtle movements of these slender appendages.

A raised whisker may signal curiosity or excitement, while a gentle flick can express annoyance or impatience. Cats, masters of non-verbal communication, use their whiskers to convey emotions and intentions with a finesse that leaves human gestures in the realm of clumsiness.

Biomimetic Applications— Unleashing the Power of Whiskers

The world of whiskers unravels the secrets of a silent language, written in the twitching tendrils of a cat's vibrissae. The wonders of whiskers extend beyond the feline realm, beckoning us to explore the incredible possibilities of biomimicry—inspiring groundbreaking advancements in robotics, sensory augmentation devices, and the very fabric of artificial intelligence.

Robotics and Navigation

Picture a world where robots effortlessly navigate complex environments, their movements guided by an unseen force akin to the graceful dance of a stalking feline. This vision is not far-fetched, thanks to the biomimetic marvel that is inspired by cat whiskers.

Engineers, drawing inspiration from the feline realm, are incorporating whisker-like sensors into robots, endowing them with an exquisite sense of touch and spatial awareness. These whisker-equipped robots can navigate darkness with finesse, avoiding obstacles and adapting to their surroundings with feline-like agility. The realm of robotics is on the cusp of a revolution, led by the silent whiskers that once prowled in the shadows.

Advancements in AI

Cat whiskers are more than just tactile sensors; they are the architects of an intricate language that conveys emotions and intentions. In the realm of artificial intelligence, scientists are unlocking the secrets of feline communication to create AI systems with unprecedented understanding and responsiveness.

Biomimetic AI, inspired by the nuanced movements of whiskers, is paving the way for machines that comprehend and respond to human cues with uncanny precision. The future of AI is not only about raw computational power; it's about the subtle art of communication, a dance inspired by the silent whispers of whiskers.

As we delve into the riveting world of biomimetic applications, the silent yet profound influence of cat whiskers beckons us to a future where the boundaries between the natural and the artificial blur. Brace yourself for a journey that transcends the ordinary, where the marvels of mammals become the catalysts for a new era of technological wonders.

Our journey through the marvels of mammals now gallops toward a majestic horizon—one adorned with the grace and power of equine wonders. Prepare to be enthralled as we delve into the tales of tails, the poetry of powerful strides, and the symphony of adaptations that make horses the epitome of grace and beauty in the animal kingdom.

Chapter 3:

Graceful Gait—

Biomimetics of a Horse's

Leg

Can a horse's leg inspire advancements in human mobility? Imagine the rhythmic dance of a thoroughbred, the effortless grace of a canter, and the sheer power unleashed with every gallop. In the mesmerizing biomechanics of a horse's leg lies a blueprint for a symphony of movement, a harmony that nature has perfected. The secrets of equine elegance invite us to explore the captivating world of biomimetics and discover how the horse's leg could hold the key to revolutionizing our own strides. Let's embark on a journey where nature's design becomes the muse for innovations that might just reshape the way we move, where the hoofbeats of inspiration echo in the corridors of human mobility.

Unraveling Equine Elegance

At first glance, the equine limb may seem like a simple assembly of bones and muscles, but beneath the surface lies a complex choreography of nature's ingenuity. Picture the legs of a horse as the dancers in an intricate ballet, each movement a harmonious blend of strength, flexibility, and precision.

The fundamental architecture of the equine limb is a testament to the artistry of these magnificent beings. From the sturdy foundation of the hooves—akin to the prima ballerina's pointe shoes—to the sinuous contours of the limbs, every element is designed for optimal function and efficiency.

The long, slender bones—metacarpals and metatarsals—provide the equine leg with both resilience and agility, allowing for swift, purposeful strides. Yet, it is the hidden marvel of the suspensory ligament, a dynamic support structure akin to a backstage crew orchestrating seamless movements, that elevates the equine gait to a level of unparalleled elegance.

Beneath the skin, the muscles form a tapestry of power, sculpted to facilitate fluid motion. The quadriceps and hamstrings, reminiscent of the principal dancers' trained muscles, work in tandem to propel the horse forward with a seemingly effortless grace. The tendons, much like taut ballet ribbons, connect muscle to bone, transmitting force with precision and finesse. As we unravel the intricate biomimetics of a horse's leg, it becomes clear that this design is not just a marvel of nature but a blueprint for innovation.

Unveiling the Equine Limb's Shock Symphony

At the heart of this biomechanical masterpiece lies a delicate balance between strength and suppleness, a duet performed by the bones, joints, and specialized structures that make up the equine limb. Picture it as a conductor leading an orchestra; every element plays its part, ensuring a harmonious response to the rhythmic percussion of hoof meeting ground.

The elegant design of the equine limb is, in essence, a shock absorber extraordinaire. The mighty bones, akin to pillars of strength, bear the brunt of each impact, distributing forces with a precision that echoes the finesse of a seasoned percussionist. Yet, it is the magical interplay of joints, tendons, and ligaments that transforms this raw power into a mesmerizing dance of adaptability.

Witness the fetlock joint—a marvel of nature's engineering—flex and extend with the fluidity of a masterful pianist's fingers. Here, the sesamoid bones act as key players, engaging in a dynamic dance that dissipates shock waves and ensures a smooth, unhindered cadence. It's a biomechanical ballet where the equine limb not only absorbs shocks but transforms them into a seamless rhythm that courses through the entire body.

You'll encounter the digital cushion, a soft and pliable structure nestled within the hoof, comparable to a cushioned dance floor that absorbs shocks with every step. This ingeniously designed mechanism enhances the comfort of the equine performer but also serves as an inspiration for innovative engineering in the world beyond the stable.

The Gallop of Green—Unleashing the Equine Limb's Energy Alchemy

In the boundless tapestry of nature's wonders, the equine limb emerges as a luminary in the realm of energy efficiency—an enchanting ballet of biomechanics designed to amplify every stride.

Picture the equine limb as a symphony conductor orchestrating a grand performance, where the efficiency of movement is the overture that captivates the audience. Beneath the sleek exterior lies an intricate web of anatomical marvels, each element finely tuned to maximize the conversion of energy, propelling the horse forward with an economy of motion that defies the limits of physics.

At the forefront of this kinetic masterpiece are the tendons, those resilient strands akin to the strings of a finely tuned instrument. As the muscles contract, they pull on these elastic connectors, storing potential energy like a bow drawn across the strings, ready to release in a crescendo of motion. This ingenious mechanism not only conserves energy but transforms it into a kinetic symphony, propelling the horse with unrivaled efficiency.

Witness the elegant dance of the tendons, ligaments, and joints, working together like a synchronized ensemble. The fetlock joint, a pivot of precision, allows for a seamless transfer of energy as the horse navigates its surroundings. With each gallop, the equine limb becomes a marvel of energy conservation, a triumph that beckons us to learn from nature's efficiency.

Unleashing Agile Robotics Through Equine Biomimetics

Imagine a world where robots seamlessly mimic the fluidity and dexterity of a thoroughbred, navigating obstacles with the grace of a seasoned equestrian. This is the essence of agile robotics—a mesmerizing fusion of equine elegance and cutting-edge technology that propels machines into a realm of adaptability previously reserved for the untamed beauty of the horse.

At the heart of this revolution lies the inspiration drawn from the intricate biomechanics of the equine limb. The principles of energy efficiency, shock absorption, and dynamic stability become the guiding beacons for engineers striving to emulate the harmonious dance of a galloping horse within the metallic sinews of a robot.

Picture a robotic limb, inspired by the supple tendons and joints of the horse, executing maneuvers with a precision that seems almost otherworldly. This agility is not a mere replication but a transformative leap, where biomimicry elevates machines beyond the rigid constraints of conventional robotics, allowing them to weave through the world with a newfound freedom.

As we delve deeper into the applications of equine biomimetics in agile robotics, we witness the birth of robotic quadrupeds—mechanical marvels that traverse landscapes with an uncanny resemblance to their living equine counterparts. These agile robots, adorned with synthetic muscles and adaptive joints, embody the fusion of nature's grace and artificial intelligence, redefining what it means to move with purpose.

Beyond mere imitation, equine biomimetics propels robots into the arena of dynamic problem-solving. The ability to adjust gait, alter stride length, and navigate diverse terrains mirrors the resourcefulness of a wild horse conquering untamed landscapes. In this marriage of biology and robotics, the agile machines of tomorrow become not just tools but partners in our quest for exploration and discovery.

Equine Biomimetics Transforms Human Mobility

Imagine a future where prosthetic limbs cease to be mere replacements and, instead, become extensions of the human body that echo the fluid grace of a thoroughbred. Step into the realm of enhanced human mobility, where equine-inspired prosthetics redefine what it means to move with both precision and poise.

At the heart of this revolution is the artful mimicry of the equine limb's biomechanics. Picture a prosthetic limb that adapts to its user's every move, mirroring the elegance and efficiency of a horse's gait. The integration of synthetic tendons, joints, and adaptive mechanisms transforms the simple act of walking into a choreographed dance—a celebration of grace, resilience, and the indomitable human spirit.

As we journey deeper, we encounter exoskeletons inspired by the equine's natural biomechanical brilliance. These wearable wonders, crafted with biomimetic finesse, amplify human strength and endurance, allowing individuals to stride beyond the limits of their physical capabilities. The symbiosis between man and machine becomes a harmonious duet, where the boundaries of what the human body can achieve are pushed further with each step.

Witness the fusion of equine biomimetics and artificial intelligence, creating prosthetic limbs that learn and adapt to their users' unique gaits. The result is not just a functional limb but a personalized companion, intuitively anticipating movement and responding with a synchronicity between the organic and the synthetic.

The Tightrope of Progress— Navigating Ethical Implications in Equine Biomimetics

Delving into the intricate dance of the horse's leg and its transformative influence on biomimetic innovation, we confront the ethical considerations that underpin this exploration. Pushing the boundaries of scientific progress while upholding ethical responsibilities requires a balanced approach, with a focus on equine research, equitable access to biomimetic advancements, and inclusive design.

As we unravel the secrets within the equine limb, we must ask ourselves: *How do we navigate this journey responsibly?* This question delves into the ethical considerations associated with equine research, emphasizing the need for researchers to act as stewards of animal welfare. It urges us to ensure that our quest for understanding doesn't overshadow the fundamental respect and care owed to our equine collaborators.

Ensuring Equitable Access to Biomimetic Advancements

The promise of biomimetic progress in prosthetics and robotics raises ethical questions about accessibility. While the potential for revolutionizing mobility is immense, we must critically examine whether the benefits are distributed fairly. It's important to consider the societal impact, questioning if the advantages of biomimetic advancements in human mobility can be extended universally. Can we build a future where these innovations benefit all individuals, irrespective of their socioeconomic status or geographic location?

Progress challenges us to navigate the ethical considerations inherent in the pursuit of equine-inspired biomimetics. As we look at the delicate balance between scientific progress and ethical responsibility, we must contemplate not only the marvels of nature but also the ethical imperative that accompanies the quest for innovation. The marvels continue, and the ethical tightrope holds the key to a future where progress and compassion walk hand in hand.

The rhythmic dance of a horse's legs is a testament to the seamless integration of form and function, prompting us to ponder the marvels hidden within the biomechanics of these majestic beings. As we continue our journey through the mesmerizing tapestry of Earth's extraordinary mammals, prepare to be captivated by the ingenious biomimetics of the elephant's trunk, where nature's engineering marvels serve as inspiration for cutting-edge technologies.

Chapter 4:

Nature's Versatile Tool—

Biomimetics of an

Elephant's Trunk

Picture a tool so versatile that it can draw water, pick up a blade of grass, uproot a tree, and even play musical notes. An elephant's trunk is more than a giant nose; it is nature's masterpiece, a marvel of design that serves a multitude of functions. In the grand orchestra of the animal kingdom, the elephant's trunk stands out as a virtuoso performer, showcasing unparalleled adaptability and ingenuity. In the realm of biomimetics, the astonishing capabilities of this natural wonder inspire human innovation and unlock the secrets to nature's versatile tool. Biological brilliance meets the boundless potential of human creativity—all through the intricacies of an elephant's trunk.

Anatomy and Function of the Trunk

The elephant's trunk stands as a testament to the ingenuity of nature, a masterpiece of muscles that seamlessly integrates strength, sensory superpowers, and prehensile precision. In the world of mammals, few adaptations are as remarkable as this multifunctional marvel that serves as a living example of nature's ability to engineer unparalleled tools for survival.

A Masterpiece of Muscles

At the core of the elephant's trunk lies an intricate network of over 40,000 muscles, making it a true masterpiece of biomechanical engineering. These muscles are arranged in multiple layers, allowing the trunk to exhibit an incredible range of motion and flexibility. The sheer number and arrangement of muscles enable the elephant to perform a myriad of tasks, from lifting and carrying heavy objects to delicately plucking the smallest leaves from a tree.

The trunk's muscular structure is not only about power but also about finesse. The muscles are capable of precise and coordinated movements, allowing the elephant to manipulate objects with incredible dexterity. This combination of strength and precision is what makes the trunk an unparalleled tool in the animal kingdom.

Sensory Superpowers

Beyond its physical prowess, the elephant's trunk is endowed with extraordinary sensory capabilities. The tip of the trunk is equipped with an impressive array of olfactory sensors, allowing elephants to detect scents over vast distances. This acute sense of smell is crucial for finding food, identifying potential threats, and locating water sources in their often vast habitats.

Additionally, the trunk serves as a sensitive tactile organ. Elephants use their trunks to explore and interact with their surroundings, conveying a wealth of

33

information about the texture, temperature, and shape of objects. This tactile acuity is particularly crucial for social interactions, as elephants use their trunks to greet each other, show affection, and establish bonds within the herd.

Fine Motor Skills and Prehensile Precision

The fine motor skills exhibited by the elephant's trunk are nothing short of extraordinary. Capable of both delicate and forceful movements, the trunk can pick up small objects, strip bark from a tree, or even draw water into the elephant's mouth for drinking. The versatility of the trunk's movements allows elephants to adapt to various ecological niches and food sources.

Furthermore, the prehensile precision of the trunk is a result of the coordinated effort of the numerous muscles and the sensory feedback it receives. Elephants can use their trunks like precision tools, whether in the careful extraction of vegetation or the manipulation of objects for problem-solving tasks. This adaptability and versatility make the elephant's trunk a true marvel of natural engineering.

Biomimetic Applications of an Elephant's Trunk

The incredible design of an elephant's trunk, with its multifunctional capabilities and versatile movements, has inspired a wave of innovation across various fields. Researchers and engineers, captivated by nature's ingenuity, have explored biomimetic applications that draw inspiration from the trunk's anatomy and function.

Robotics and Prosthetics

In the realm of robotics, the biomimetic approach to designing robotic arms and manipulators has been significantly influenced by the elephant's trunk. Engineers have sought to replicate the muscular structure and flexibility of the trunk to create robotic appendages that can perform a wide range of tasks with precision and adaptability.

These biomimetic robots find applications in environments where traditional rigid robots might struggle, such as navigating through cluttered spaces or interacting delicately with objects. Additionally, the fine motor skills and sensory capabilities of the elephant's trunk have informed the development of robotic prosthetics, offering a more natural and intuitive interface for amputees.

Industrial Manipulators

Industries that require precision manipulation of objects in confined spaces have turned to biomimicry inspired by the elephant's trunk. Industrial manipulators designed with a flexible and muscular structure can navigate complex environments, mimicking the trunk's ability to reach and handle objects with finesse.

In manufacturing and assembly lines, these biomimetic manipulators enhance efficiency and adaptability, allowing for more agile and versatile operations. The trunk-inspired industrial manipulators also excel in tasks that demand both strength and precision, showcasing the advantages of nature-inspired design in enhancing technological capabilities.

Medical Advancements

The biomimetic applications of the elephant's trunk extend to the field of medicine, where researchers draw inspiration for innovative medical devices and surgical tools. The trunk's combination of strength, flexibility, and sensory feedback has influenced the design of minimally invasive surgical instruments, enabling surgeons to perform delicate procedures with enhanced precision.

Furthermore, the trunk's sensory superpowers have inspired the development of advanced diagnostic tools. Researchers explore incorporating sensitive tactile sensors into medical devices to enhance the detection and characterization of tissues during medical procedures. The goal is to improve the accuracy of diagnoses and treatments by leveraging the trunk's natural ability to gather detailed information about its surroundings.

As scientists continue to unravel the mysteries of nature's versatile tool, the integration of biomimicry into various fields promises to shape a future where technology seamlessly mirrors the efficiency and adaptability found in the remarkable wonders of the animal kingdom.

Ethical Considerations and Conservation

The biomimetic exploration of the elephant's trunk is not only a journey into the realms of technological innovation but also an opportunity to reflect on the ethical considerations surrounding the conservation of these majestic creatures. As we draw inspiration from nature's versatile tool, it is crucial to address the challenges faced by elephants and ensure that our endeavors contribute to their well-being.

The Plight of Elephants

Before delving into the ethical considerations, it's essential to acknowledge the dire situation facing elephants in the wild. These intelligent and social

As we delve into the extraordinary world of mammals and explore the wonders of nature's versatile tools, our journey takes an intriguing turn. While the elephant's trunk captivates us with its biomechanical brilliance, pangolins present another marvel, a testament to nature's endless creativity in bestowing extraordinary abilities upon its creatures. Let's uncover the secrets of these scaly wonders and unravel the biomimetic potential they offer to the world of technological innovation.

Chapter 5:

Armored Marvels—

Biomimetics of Pangolins

Within Earth's biodiversity, certain creatures stand out as marvels of intelligence and resilience. Among these, pangolins emerge as truly extraordinary beings, captivating the imagination with their enigmatic charm and the remarkable armor that adorns their bodies. The unique construction of their scales has captivated scientists, engineers, and researchers seeking inspiration for cutting-edge technologies.

Often referred to as "scaly anteaters," pangolins belong to the order Pholidota and are the only mammals equipped with protective keratin scales covering their bodies. This armor showcases incredible versatility in the face of diverse environments—from the dense rainforests of Southeast Asia to the arid landscapes of Africa. The story of these armored marvels is not only a testament to the wonders of nature but also a source of inspiration for human ingenuity, sparking a new era of innovation influenced by the fascinating biomimetics of pangolins.

Pangolin Anatomy and Behavior

The armor of pangolins unveils the potential applications of biomimicry in various fields, from materials science to robotics. The intricate design of their scales, which overlap like a natural suit of armor, offers insights into creating lightweight, durable materials that can withstand extreme conditions. Moreover, the way pangolins move, navigate, and protect themselves inspires innovations in robotics, paving the way for the development of more efficient and adaptable machines. Armed with a unique suit of armor, these enigmatic mammals navigate their ecosystems with a grace that belies their tough exterior.

Let's unravel the mysteries of pangolin anatomy and behavior, shedding light on these remarkable armored marvels of the mammalian world.

Scales as Defensive Mechanisms

At the heart of the pangolin's allure lies its extraordinary armor—scales that not only define its appearance but also serve as an impenetrable defense mechanism. Composed primarily of keratin, the same protein found in human hair and nails, these scales form overlapping patterns that create a formidable shield against potential predators.

The arrangement of these scales is crucial to the pangolin's defense. When threatened, the pangolin instinctively curls into a tight ball, presenting an impenetrable surface of scales to any would-be assailant. This effectively protects the pangolin's vulnerable underbelly, creating a biological shield.

What makes these scales even more fascinating is their regenerative ability. Pangolins can shed and regrow their scales throughout their lives, ensuring that their defensive armor remains intact. This regenerative capability not only aids in healing wounds but also presents a potential avenue for biomimetic research in the field of materials science.

Digestive Adaptations

While the exterior of pangolins showcases a fortress-like defense, their internal anatomy reveals an equally fascinating story. Pangolins are primarily insectivorous, feasting on ants and termites with their long, sticky tongues. However, their digestive system sets them apart from other mammals and contributes to their classification as true marvels of nature.

One of the most distinctive features of pangolin anatomy is the absence of teeth in their mouths. Instead, they rely on a muscular stomach with keratinous spines to grind and break down the tough exoskeletons of their insect prey. This allows pangolins to efficiently extract nutrients from their diet, compensating for the lack of conventional chewing mechanisms.

The stomach's unique structure also serves as a potential area for biomimetic exploration. Researchers are intrigued by the efficiency of the pangolin's digestive system and are exploring ways to incorporate similar principles into waste management technologies and industrial processes. The ability to break down tough materials could revolutionize waste disposal methods and contribute to a more sustainable future.

Beyond Nature's Marvel

The exploration of pangolin anatomy and behavior reveals a story of resilience and unique abilities with potential for biomimicry. From the intricate design of their defensive armor to the efficiency of their digestive system, pangolins offer a wealth of inspiration for scientific innovation. The natural capabilities of these armored marvels offer a wealth of inspiration for cutting-edge innovation through biomimetics.

Protective Armor Innovations

The armor of pangolins has long captivated the imaginations of scientists and engineers alike. The intricate patterns and overlapping scales form a defense mechanism that combines strength, flexibility, and regenerative capabilities—qualities that have spurred interest in various fields of materials science.

Researchers are delving into the microstructure of pangolin scales, seeking to replicate their design for practical applications. The unique way in which the scales overlap provides lightweight yet robust protection, offering potential breakthroughs in developing advanced materials for protective gear, armor, and even flexible electronics.

One promising avenue is the creation of biomimetic textiles inspired by the pangolin's armor. By mimicking the scale patterns and the flexibility of the armor, scientists aim to produce lightweight, breathable, and highly durable textiles. These textiles could find applications in diverse fields, from sports equipment and military gear to spacesuits and medical devices.

Moreover, the regenerative capabilities of pangolin scales are an area of particular interest. Researchers are exploring how these principles could be applied to create self-healing materials. Imagine a future where buildings, vehicles, and infrastructure possess the ability to repair themselves, inspired by the regenerative prowess of the pangolin's protective armor.

Pharmaceutical Potential

Beyond their defensive armor, pangolins harbor another secret that holds immense promise for the field of pharmaceuticals. Traditional Chinese medicine has long sought pangolin scales for their perceived medicinal properties, but it is the potential within their biological makeup that truly captivates the scientific community.

Pangolins have a unique immune system that enables them to coexist with a variety of pathogens, including those that are deadly to humans. Understanding the intricacies of their immune response may hold the key to developing innovative antiviral medications and vaccines.

The study of pangolin immune systems has gained particular importance in the wake of zoonotic diseases, where pathogens jump from animals to humans. The recent spotlight on pangolins as potential intermediate hosts for coronaviruses highlights the urgency of understanding their immune system and applying this knowledge to mitigate the risk of future pandemics.

Biomimicry in this context involves deciphering the molecular and cellular mechanisms of the pangolin immune system to develop pharmaceutical interventions. Researchers are exploring the possibility of creating synthetic compounds inspired by the unique features of pangolin immunity, potentially leading to breakthroughs in antiviral drugs and vaccines.

Ethical Considerations and Wildlife Conservation

In our exploration of armored marvels and biomimetics inspired by pangolins, it is crucial to confront the ethical considerations that arise from our fascination with these creatures. The threat to pangolins, the ethical use of pangolin-inspired designs, and conservation through biomimicry offer insights into the delicate balance between scientific innovation and wildlife conservation.

The Threat to Pangolins

Pangolins, despite their extraordinary adaptations, find themselves at the center of a perilous situation. The demand for their scales and body parts, fueled by traditional beliefs in some cultures, has led to a devastating surge in illegal wildlife trade. This, coupled with habitat loss and other anthropogenic pressures, has pushed pangolin species to the brink of extinction.

As we draw inspiration from pangolins for biomimetic designs, it becomes imperative to recognize and address the threats these creatures face. The ethical pursuit of knowledge and innovation demands a simultaneous commitment to the preservation of pangolin populations and their habitats.

Conservation efforts must extend beyond biomimicry applications, encompassing strategies to combat illegal wildlife trade, protect natural habitats, and raise awareness about the ecological importance of pangolins. By acknowledging the ethical imperative of safeguarding these creatures, we ensure that our exploration of pangolin-inspired biomimetics is grounded in a responsible and conscientious approach.

Ethical Use of Pangolin-Inspired Designs

The marvels of pangolin biology offer a wellspring of inspiration for scientific and technological advancements. However, the ethical use of pangolin-inspired designs necessitates a thoughtful and considerate approach. Biomimicry, while offering innovative solutions, should not inadvertently contribute to the further decline of pangolin populations or harm the ecosystems they inhabit.

animals are confronted with threats such as habitat loss, poaching for ivory, and human-wildlife conflict. It is imperative to recognize our responsibility in safeguarding these incredible creatures from extinction.

Conservation efforts must focus on protecting elephant habitats, enforcing anti-poaching measures, and promoting coexistence between humans and elephants. Acknowledging and addressing the challenges faced by elephants forms the foundation for ethical biomimicry inspired by these magnificent beings.

Responsible Biomimicry

As scientists and engineers explore biomimetic applications inspired by the elephant's trunk, ethical considerations become paramount. Responsible biomimicry involves not only acknowledging the source of inspiration but also actively contributing to the conservation of the species in question. It requires a commitment to sustainability, environmental responsibility, and respect for the natural world.

Researchers must prioritize sourcing materials ethically, minimizing environmental impact, and ensuring that biomimetic technologies do not inadvertently contribute to the exploitation of wildlife or their habitats. By adopting responsible biomimicry practices, we can create a symbiotic relationship between technological innovation and the conservation of the animals that inspire it.

In the realm of materials science, where the structure of pangolin scales is being investigated for its potential applications, ethical considerations are paramount. Researchers must prioritize sustainable practices and ensure that their work does not inadvertently encourage the illegal trade of pangolin scales. Collaboration with conservation organizations, adherence to ethical guidelines, and transparency in research practices become essential pillars of responsible biomimetic exploration.

Moreover, the development of pharmaceuticals inspired by pangolin immune systems should prioritize ethical sourcing of biological materials. The principles of fair and equitable access to genetic resources, as outlined in international agreements such as the Nagoya Protocol, should guide researchers in their quest to harness the therapeutic potential of pangolin-inspired biomimicry.

A holistic approach to ethics in biomimetics involves not only avoiding harm but actively contributing to the well-being of pangolin populations. This may include supporting conservation initiatives, participating in community-based projects, and fostering awareness about the importance of pangolins in their ecosystems.

Ultimately, the ethical considerations surrounding biomimetics and wildlife conservation underscore the responsibility we bear as stewards of the natural world. In embracing this responsibility, we embark on a journey where the marvels of mammals, the wonders of biomimicry, and the imperative of ethical stewardship coalesce into a vision of a sustainable and interconnected future.

Conservation Through Biomimetics

While pangolins face imminent threats from illegal wildlife trade and habitat loss, biomimetics emerges as a powerful tool in the conservation toolbox. Inspired by the unique characteristics of pangolins, scientists are devising innovative strategies to protect these armored marvels and their ecosystems.

One such initiative involves the development of biomimetic sensors to combat wildlife trafficking. Drawing inspiration from the extraordinary olfactory senses of pangolins, researchers are working on creating artificial noses that can detect the scent of pangolin scales, aiding law enforcement agencies in intercepting illegal shipments. This application not only protects pangolins directly but also contributes to the broader fight against wildlife crime.

Biomimicry also plays a role in habitat restoration efforts. By studying the ecological impact of pangolins in their native environments, conservationists gain insights into the intricate web of relationships within ecosystems. Mimicking these natural processes can inform sustainable practices for land use, contributing to the conservation of biodiversity beyond pangolin species.

Moreover, biomimetic approaches can be employed to develop alternative livelihoods for communities in proximity to pangolin habitats. By drawing inspiration from the way pangolins contribute to pest control, researchers explore sustainable agricultural practices that balance human needs with the preservation of natural ecosystems.

In essence, conservation through biomimetics involves not only protecting pangolins directly but also leveraging their unique traits to foster a harmonious coexistence between humans and nature. The journey into the biomimetic realm of pangolins is not just a scientific exploration but a call to action. It beckons us to appreciate the intricate balance of nature, to draw inspiration responsibly, and to weave the marvels of mammals into the fabric of human progress.

From the armored pangolins, we plunge into the Arctic world of ground squirrels, where the marvels of mammals take on an entirely different guise. As we journey from the tropical landscapes inhabited by pangolins to the frigid expanses of the Arctic, the diversity of nature's innovations unfolds before us.

Chapter 6:

Surviving the Chill—

Biomimetics of Arctic

Ground Squirrels

In the hushed landscape of the Arctic tundra, where frigid winds sweep across vast expanses of snow and ice, a remarkable creature thrives against all odds. The Arctic ground squirrel, a master of survival in one of the world's harshest environments, holds the key to unlocking the secrets of winter endurance. Let's delve into the extraordinary world of these resilient mammals and uncover the ingenious strategies they employ to conquer the bone-chilling cold.

Nature has equipped the Arctic ground squirrel with an arsenal of survival mechanisms that not only defy the harsh Arctic conditions but also inspire innovations in the field of biomimetics. These remarkable squirrels navigate the extreme challenges of winter with hibernation strategies that seem almost otherworldly and unique biological capabilities that enable them to endure temperatures that would freeze the most resilient of beings. Prepare to be amazed by the biomimetics of Arctic ground squirrels, where nature's ingenuity meets human innovation.

Capacity for Extreme Cold

In the Arctic, where temperatures plummet to levels that would challenge even the hardiest of beings, the Arctic ground squirrel emerges as a true marvel of mammals. To understand the secrets behind their winter survival, we must unravel the intricate tapestry of their adaptations for extreme cold—a saga that encompasses the wonders of hibernation and torpor, the strategic employment of supercooling, and the fascinating mechanisms employed for energy conservation and mitigating brain neuron decay.

Hibernation and Torpor—Mastering the Art of Winter Slumber

One of the Arctic ground squirrel's most captivating abilities is to enter a state of suspended animation—hibernation. As winter descends upon the Arctic landscape, these squirrels retreat to their burrows, where they experience a profound drop in body temperature, heart rate, and metabolic activity. This orchestrated slowdown of physiological functions allows them to endure the scarcity of food and the brutal cold with remarkable efficiency.

During hibernation, the squirrel's body temperature can plummet to near-freezing levels, hovering just above 32 °F (0 °C), a temperature that would spell certain death for most mammals. Yet, in this seemingly lifeless state, the squirrel conserves energy at an astonishing rate, minimizing the need for vital resources during the harsh winter months.

Torpor, a state of temporary metabolic reduction, further enhances their survival strategy. In torpor, the squirrel periodically arouses from deep hibernation to raise its body temperature and engage in essential activities such as grooming and excretion. This cyclical pattern allows them to reap the benefits of hibernation while intermittently addressing biological necessities.

Scientists studying these remarkable instincts in squirrels are actively exploring ways to harness the power of torpor for human applications, such as in the field of space travel where long-duration missions require innovative solutions for resource conservation and energy efficiency.

Supercooling Strategies—Antifreeze for Arctic Squirrels

Surviving in the Arctic requires more than just a tolerance for cold; it demands a defense against the very freezing of bodily fluids. The squirrel has a unique antifreeze strategy known as supercooling.

While many mammals risk the formation of ice crystals within their bodies at sub-zero temperatures, these resourceful squirrels can lower their body temperature without freezing solid. By carefully managing the freezing point of their bodily fluids, they prevent ice crystals from forming and causing cellular damage.

Intriguingly, the squirrel's blood can remain in a liquid state at temperatures as low as - 26.8 °F (2.9 °C), allowing them to endure the coldest Arctic nights without the detrimental effects of freezing. This remarkable capability has inspired researchers in the development of antifreeze agents for medical and industrial applications, where preventing the formation of ice crystals is crucial.

Energy Conservation and Brain Neuron Decay—Navigating the Winter Cognitive Challenge

Surviving the Arctic winter is not solely a battle against the elements; it is also a test of cognitive endurance. The Arctic ground squirrel faces the challenge of maintaining brain function during extended periods of hibernation, where metabolic activity is drastically reduced.

To cope with this challenge, these squirrels have a unique strategy to mitigate the effects of neuron decay. During hibernation, the Arctic ground squirrel experiences a significant decrease in brain temperature, allowing for reduced metabolic activity and minimizing the energy demands on neural tissues. This extraordinary feat of biological engineering enables the squirrel to sustain cognitive function over prolonged periods of low metabolic activity.

Researchers investigating the neuroprotective mechanisms employed by Arctic ground squirrels have identified potential applications in fields such as medicine, where preserving brain function during periods of reduced metabolic activity could have profound implications for conditions such as stroke and traumatic brain injury.

From the artful dance of hibernation and torpor to the antifreeze prowess of supercooling and the strategic conservation of cognitive function, these adaptations not only ensure survival in the Arctic wilderness but also inspire groundbreaking innovations in the field of biomimetics.

The Remarkable Intervals of Awakening

Within the frozen realms of the Arctic tundra, the Arctic ground squirrel weaves a narrative of survival that transcends the ordinary. As we explore the intricacies of their hibernation, a fascinating phenomenon emerges: every two or three weeks, these resilient creatures break the icy silence with a deliberate act of defiance against the relentless cold. Amazingly, the Arctic ground squirrel shivers itself back to its normal body temperature of 97.5 °F (36.4 °C), marking a brief but crucial interval that holds the key to the brain's survival and remaining unaffected by the numbing chill for a remarkable span of 12 to 15 hours. In this brief but intense burst of metabolic activity, the squirrel temporarily emerges from the frigid embrace of hibernation.

The Brain's Winter Symphony—Enhancing Cognitive Resilience

Researchers probing the intricacies of this awakening phenomenon have uncovered a fascinating revelation—this brief warming period plays a crucial role in the brain's survival. Despite the challenges posed by hibernation, the Arctic ground squirrel ensures that its most vital organ, the brain, is not left in the cold. This periodic awakening is believed to serve as a strategic intervention, preventing prolonged exposure to suboptimal temperatures that could compromise the delicate balance of neural functions.

One study has raised eyebrows in scientific circles. It suggests that not only does the squirrel's brain survive the harsh winter conditions, but it may actually function even better after the period of hibernation (Jabr, 2012). The intricacies of this phenomenon are still being unraveled, but it appears that the unique interplay of metabolic activity and temperature regulation during the awakening intervals contributes to an enhanced cognitive state.

Biomimetics in Action—Lessons for Human Innovation

The Arctic ground squirrel's ability to momentarily break free from the grip of hibernation, revive its body temperature, and safeguard the brain's functionality provides a blueprint for developing technologies that optimize performance under extreme conditions. From medical applications that aim to preserve cognitive function during periods of reduced activity to advancements in temperature regulation systems for critical environments, the biomimetics of the Arctic ground squirrel opens doors to a realm of possibilities for human progress.

Brain Loss and Regenerative Abilities— Lessons From the Arctic Survivors

One of the most astonishing facets of the Arctic ground squirrel lies in its regenerative abilities, especially concerning the brain. As these resilient creatures navigate the challenges of hibernation, during which they experience a significant reduction in metabolic activity and body temperature, they manage to safeguard their brains from the detrimental effects of prolonged dormancy.

The periodic intervals of awakening, during which the Arctic ground squirrel shivers itself back to its normal body temperature, play a pivotal role in mitigating potential brain damage. Researchers have been astounded to discover that despite the fluctuations in metabolic activity and the extreme conditions of hibernation, the squirrel's brain not only endures but shows signs of regenerative potential.

This regenerative prowess challenges conventional notions about the limitations of mammalian brain regeneration. While human brains are often considered to have limited regenerative abilities, the Arctic ground squirrel's resilience suggests that there might be untapped potential for regrowth and repair that transcends our current understanding.

Biomedical Insights for Treating Memory Loss in Alzheimer's Patients

The parallels between the Arctic ground squirrel's brain resilience and the challenges faced by humans suffering from neurodegenerative diseases, such as Alzheimer's, are striking. Alzheimer's disease, characterized by progressive memory loss and cognitive decline, poses a significant burden on individuals and their families. The search for effective treatments and preventive measures has become a focal point in biomedical research.

The periodic rewarming intervals during the Arctic ground squirrel's hibernation offer valuable insights into addressing memory loss. As researchers delve into the mechanisms behind the squirrel's ability to protect and potentially regenerate brain cells, they explore avenues that could revolutionize the understanding and treatment of Alzheimer's disease.

One study suggests that the Arctic ground squirrel's brain experiences enhanced cognitive function after hibernation (Drew et al., 2016). This phenomenon sparks hope that the regenerative mechanisms employed by these remarkable mammals could inform strategies for preventing or even reversing the cellular damage associated with Alzheimer's and other neurodegenerative conditions.

Unlocking the Brain's Potential—A Biomedical Odyssey

The study of the Arctic ground squirrel holds a transformative potential for unlocking the mysteries of the human brain's resilience and regenerative capacities. Researchers aim to decipher the intricate biological processes that enable these mammals to endure extreme conditions without compromising cognitive function. As scientists unravel the secrets embedded in the Arctic ground squirrel's neuroprotective mechanisms, they envision a future where biomedical interventions harness the regenerative potential of the human brain.

Biomimetics in the Service of Human Health

The applications of Arctic ground squirrel-inspired insights in the field of biomedical research extend beyond Alzheimer's disease. Researchers are exploring the potential of these adaptations for addressing a spectrum of neurological disorders and brain injuries. The ability to induce controlled states of reduced metabolic activity, mimicking aspects of hibernation, holds promise for preserving brain function during critical periods, such as following traumatic brain injury or stroke.

Additionally, the regenerative potential observed in the Arctic ground squirrel's brain prompts investigations into strategies for enhancing neurogenesis in humans, offering hope for the development of innovative therapies for various neurological conditions.

Hope on the Horizon—Preventing and Reversing Cellular Damage

In the pursuit of understanding the Arctic ground squirrel's brain wonders, the ultimate goal for researchers is to chart a course toward preventing or even reversing cellular damage in the human brain. The regenerative capacities hinted at by these Arctic survivors inspire a vision of a future where neurodegenerative diseases are not just managed but actively treated, and where the resilience of the human brain becomes a beacon of hope. The biomedical applications of these insights open avenues for groundbreaking advancements, shaping a future where the marvels of mammals serve as catalysts for transformative breakthroughs in human health and well-being.

Ethical Considerations and Climate Action

In the pursuit of unlocking the marvels of Arctic ground squirrels and translating these into potential solutions through biomimetics, it is imperative to navigate the ethical dimensions that accompany such endeavors. From the impact of climate change on Arctic habitats to the ethical use of knowledge gained from brain memory research, it's important to delve into the considerations that guide our exploration of the marvels of mammals in the context of climate action and ethical responsibility.

Climate Change and Arctic Habitats—A Threat to Marvels of Resilience

The Arctic, a vast and delicate ecosystem, is undergoing unprecedented changes due to the effects of climate change. As temperatures rise, ice melts, and habitats transform, the very survival of Arctic species, including the resilient Arctic ground squirrel, hangs in the balance.

Understanding the marvels of Arctic ground squirrels becomes intertwined with the urgent need for climate action. As we explore their capability for surviving the chill, it becomes clear that these abilities are in harmony with the delicate balance of the Arctic environment. Ethical considerations compel us to recognize the interconnectedness of all living beings and to address climate change as a global responsibility.

Biomimetics inspired by Arctic ground squirrels must not be divorced from the broader context of environmental stewardship. The ethical imperative is to leverage this knowledge not just for technological innovation but also for fostering a deeper understanding of our role in preserving the habitats that nurture these marvels of resilience.

Ethical Use for Brain Memory Research— Balancing Curiosity and Caution

The insights gained from studying the Arctic ground squirrel's brain hold immense potential for addressing memory loss and neurodegenerative diseases. However, ethical considerations must guide the ethical use of this knowledge to ensure that scientific curiosity does not compromise principles of respect, consent, and responsible research.

The squirrel's brain is not merely a trove of data to be exploited but a testament to the intricate beauty of nature's designs. The ethical use of this knowledge involves respecting the autonomy and dignity of the subjects, whether they are human participants in studies inspired by squirrel research or the animals themselves.

Moreover, ethical considerations extend to how the knowledge gained from this research is disseminated and applied. Transparency, collaboration, and a commitment to the well-being of both human and animal subjects must guide the ethical trajectory of biomimetic research inspired by Arctic ground squirrels.

From the icy north, we now take flight, soaring into the nocturnal realm with bats and their exceptional echolocation wonders. As we transition from the frost-kissed landscapes of the Arctic to the darkness cloaked in the secrets of the night, the marvels of mammals continue to unfold, each species weaving its unique tale of expertise and resilience. We now delve into the mysteries of the sky, exploring how bats, with their extraordinary echolocation abilities, navigate the darkness with unparalleled precision, echoing the marvels that nature has crafted.

Chapter 7:

Sonic Navigators—

Biomimetics of Bats'

Echolocation

In the inky embrace of the night, where darkness reigns supreme, bats emerge as nature's unparalleled aviators, navigating through obscurity with a precision that borders on the supernatural. At the heart of their nocturnal prowess lies a phenomenon as intricate as it is astonishing: echolocation. This sophisticated ability, rooted in the realm of bioacoustics, empowers bats to perceive their surroundings with unparalleled acuity, making them the masters of the sonic world.

The remarkable skill of echolocation allows bats to seamlessly maneuver through the obsidian tapestry of the night. Let's embark on a journey to unravel the secrets behind the bio-inspired marvel of bat echolocation, unveiling the incredible biomimetics of bats and exploring how these winged wonders harness the power of sound waves to not only survive but thrive in the mysterious realm of the night.

Bioacoustics of Bats

To comprehend the science of echolocation, we must first delve into the captivating realm of bioacoustics—the study of how animals use sound for communication, navigation, and other ecological functions. For bats, this translates into an extraordinary talent that hinges on their ability to emit and interpret high-frequency sound waves.

Bats emit ultrasonic pulses, well beyond the range of human hearing, typically between 20 and 200 kilohertz. These pulses act as sonic probes, sent out into the darkness to interact with the environment. As the sound waves encounter obstacles or prey, they bounce back as echoes, forming a detailed auditory map that bats skillfully interpret in real time.

The sophistication of bat echolocation lies not only in the emission of these ultrasonic pulses but also in their ability to adjust the frequency, duration, and intensity of these calls based on the specific requirements of their surroundings. Different bat species have distinct echolocation tailored to their ecological niches and hunting strategies.

High-Frequency Sound Production

Central to the marvel of bat echolocation is the generation of ultrasonic sound waves, a feat accomplished through the anatomical wonders of the larynx, or vocal folds. Unlike human vocal cords, which produce audible sounds, the vocal folds of bats are specialized for emitting ultrasonic frequencies.

Within the larynx, bats possess a unique set of muscles and ligaments that allow them to manipulate the tension and length of their vocal folds rapidly. This enables them to produce ultrasonic calls with remarkable precision. Some bat species can emit calls at rates exceeding 200 per second, creating a barrage of sound waves that saturate the environment and provide a detailed auditory snapshot.

The diversity in echolocation strategies among different bat species is astounding. Some bats emit constant-frequency calls, maintaining a consistent pitch throughout, while others modulate the frequency, creating a dynamic and adaptable sonar system. This variability in echolocation techniques reflects the intricate interplay between the physiology of bats and the ecological challenges they face in their respective habitats.

Sound Reception and Interpretation

Echolocation extends beyond mere sound emission; equally crucial is the bats' ability to receive and interpret the returning echoes. The ears of bats are finely tuned instruments, capable of capturing and processing ultrasonic signals with astonishing precision.

Most bats have large, highly sensitive ears that can detect a wide range of frequencies, allowing them to pick up even the faintest echoes. In some species, the ears are asymmetrical, providing bats with the ability to discern the directionality of sound, a crucial skill for pinpointing the location of prey or obstacles in the dark.

As bats fly through their surroundings, they continuously process the incoming echoes in their brains. The time delay, frequency shift, and intensity of these echoes convey a wealth of information about the size, shape, distance, and texture of objects in their path. It's akin to a three-dimensional auditory hologram, allowing bats to construct a detailed mental map of their environment on the fly.

The neural complexity involved in this real-time interpretation of echoes is nothing short of astonishing. Bats possess specialized brain regions dedicated to processing echolocation information, allowing them to make split-second decisions as they navigate through cluttered environments, pursue elusive prey, or avoid potential hazards.

The Advantage of Echolocation

Echolocation has conferred upon bats a remarkable advantage, enabling them to exploit ecological niches that would be challenging or inaccessible to other nocturnal creatures. From open spaces to dense forests, bats have demonstrated unparalleled versatility in deploying their echolocation skills across diverse habitats.

In open environments, bats can extend their range of echolocation to detect prey or obstacles over considerable distances. Conversely, in cluttered habitats like dense forests, where echoes can bounce off multiple surfaces, bats showcase an ability to use their echolocation calls to effectively navigate through complex and dynamic environments.

Beyond the realm of navigation, echolocation serves as a potent tool for hunting. Bats can detect and track the slightest movements of insects, often capturing them mid-flight with astounding precision. This hunting prowess has positioned bats as vital contributors to pest control, with many species consuming vast quantities of agricultural pests and disease-carrying mosquitoes.

Future Implications and Inspirations

The intricate science of bat echolocation not only fascinates researchers studying animal behavior and physiology but also holds considerable promise for human innovation. Biomimicry has already drawn inspiration from bat echolocation for various technological applications.

Sonar technology, used in submarines and underwater navigation, has borrowed concepts from bat echolocation to improve detection capabilities in challenging acoustic environments. Robotics and drone design have also benefited from insights derived from the agility and precision of bat flight and echolocation, paving the way for more efficient and adaptable autonomous systems.

As we unravel the intricacies of bat echolocation, we find ourselves not only in awe of the natural world but also inspired by the potential applications that emerge from understanding and emulating these marvelous abilities.

Biomimetic Sonar and Acoustic Technology—Unleashing the Power of Bat-Inspired Innovation

The marvels of bat echolocation transcend the boundaries of the natural world, inspiring a wave of technological breakthroughs in the form of biomimetic sonar and acoustic technology. Drawing inspiration from the intricate mechanisms of bat echolocation, scientists and engineers have harnessed the power of sound to revolutionize various fields, from navigation and mapping innovations to transformative applications in medical imaging and assistance devices for the visually impaired.

Navigation and Mapping Innovations

The rise of biomimetic sonar has significantly impacted the world of navigation, offering solutions that surpass traditional methods and extend into previously uncharted territories. Taking cues from bats, whose echolocation abilities enable them to navigate through complex environments with unmatched precision, engineers have developed sonar systems that enhance navigation and mapping in both terrestrial and aquatic environments.

Underwater exploration has seen a particular transformation, with biomimetic sonar systems being employed in autonomous underwater vehicles (AUVs) and submarines. These systems emulate the echolocation capabilities of bats to navigate through challenging underwater terrains, avoiding obstacles and mapping the topography with remarkable accuracy. The adaptability of biomimetic sonar in aquatic environments is particularly noteworthy, as it provides a dynamic solution for tasks ranging from environmental monitoring to search and rescue operations.

In aerial applications, biomimetic sonar-equipped drones and unmanned aerial vehicles (UAVs) have demonstrated enhanced obstacle avoidance and navigation capabilities. By replicating the echolocation principles of bats, these airborne systems can navigate through dense foliage, urban landscapes, and other complex environments with a precision that goes beyond traditional visual-based navigation methods.

Medical Imaging Applications

The transformative potential of biomimetic sonar extends beyond the realms of navigation, seeping into the domain of medical imaging with groundbreaking applications. Inspired by the intricate biological processes underlying bat echolocation, researchers have developed innovative sonar-based medical imaging techniques that offer a non-invasive and highly detailed view of the human body.

One of the most promising applications lies in the realm of ultrasound imaging. By adopting principles from bat echolocation, ultrasound technology has advanced to provide clearer and more detailed images for various medical diagnostics. The ability to visualize internal structures with high resolution and in real-time has revolutionized prenatal care, cardiovascular diagnostics, and musculoskeletal imaging, among other medical fields.

Moreover, biomimetic sonar has played a pivotal role in the development of advanced imaging modalities, such as photoacoustic imaging. This technique combines laser-induced ultrasound with traditional imaging methods, allowing for deep tissue penetration and precise visualization of anatomical structures and abnormalities. The integration of bat-inspired sonar principles into medical imaging technologies continues to push the boundaries of what is possible in non-invasive diagnostics and medical research.

Advancements in Hearing Aids and Electronic Canes for the Blind

The impact of biomimetic sonar reaches beyond the confines of laboratories and medical facilities, profoundly influencing the lives of individuals with sensory impairments. Inspired by the natural echolocation abilities of bats, researchers have explored innovative applications in hearing aids and electronic canes for the blind, ushering in a new era of assistive technologies.

In the realm of hearing aids, biomimetic sonar has contributed to the development of devices that not only amplify sounds but also enhance spatial awareness. By incorporating principles akin to bat echolocation, these advanced hearing aids can selectively focus on specific sounds and suppress background noise, providing users with a more natural and immersive auditory experience. The result is a technology that adapts to the user's environment, mimicking the adaptability of bats in different acoustic settings.

For individuals with visual impairments, electronic canes enhanced by biomimetic sonar have emerged as powerful tools for navigation and obstacle avoidance. These canes are equipped with ultrasonic sensors that emit and detect sound waves, creating a virtual map of the surroundings. As obstacles or potential hazards are encountered, the cane provides haptic or auditory feedback, allowing users to navigate through complex environments with increased confidence and independence.

Future Horizons—From Inspiration to Innovation

The integration of biomimetic sonar and acoustic technology into various facets of our lives marks a testament to the power of nature-inspired innovation. The fields of robotics, autonomous systems, and artificial intelligence stand to benefit significantly from the principles derived from bat-inspired sonar. As we continue to unravel the mysteries of bat echolocation, new possibilities emerge on the horizon, holding the potential to reshape industries and improve the quality of life for countless individuals.

Echoes of Ethics—Navigating the Intersection of Biomimetic Sonar and Wildlife Conservation

The marvels of bats' echolocation take center stage, captivating the human imagination and inspiring groundbreaking technologies. However, as we delve into the realm of biomimetics and the ethical implications it carries, it becomes imperative to navigate the delicate balance between innovation and the preservation of our natural world.

Bats and Ecosystem Health

Before delving into the ethical considerations, it is crucial to acknowledge the pivotal role that bats play in maintaining ecosystem health. As natural stewards of the night, bats contribute significantly to insect control, pollination, and seed dispersal. Their ecological services are invaluable, making them indispensable components of diverse ecosystems worldwide.

However, bats face numerous threats, including habitat loss, climate change, and the spread of diseases such as white-nose syndrome. Understanding the ethical implications of biomimetic sonar requires a holistic appreciation of bats' ecological contributions and the potential impact of human activities, both positive and negative, on these remarkable creatures.

When developing echolocation-inspired technologies, a primary ethical consideration lies in ensuring that innovation does not inadvertently harm bat populations or disrupt their vital ecological roles. Striking a balance between technological progress and wildlife conservation becomes a paramount objective, fostering a harmonious coexistence between human innovation and the preservation of biodiversity.

Ethical Use of Echolocation-Inspired Technologies

As biomimetic sonar technologies find their way into various applications, ethical considerations become a guiding compass in their development, deployment, and utilization. The responsible use of echolocation-inspired technologies involves a multi-faceted approach that prioritizes ecological integrity, animal welfare, and the preservation of biodiversity.

Establishing collaborative partnerships with wildlife conservation organizations is essential for ethical biomimetic sonar development. Conservation experts can provide invaluable insights into local ecosystems, help identify potential risks, and contribute to the creation of guidelines that prioritize ecological preservation. This collaborative approach ensures that technological advancements align with broader conservation goals.

Ethical innovation also necessitates ongoing research and monitoring to assess the impact of biomimetic sonar technologies on wildlife. This includes studying the potential effects on bat behavior, migratory patterns, and reproductive success. By fostering a continuous assessment and improvement culture, developers can adapt their technologies to align with evolving ethical standards and ecological understanding.

Mitigating Environmental Impact

Ethical considerations begin with minimizing the potential environmental impact of biomimetic sonar systems. Whether employed in navigation, medical imaging, or assistive technologies, developers must prioritize designs that operate within environmentally sustainable parameters. This involves ensuring that the energy requirements, emissions, and waste associated with these technologies are minimized to prevent adverse effects on ecosystems.

Wildlife-Friendly Design

Designing technologies that respect the natural behaviors of wildlife is a key ethical consideration. For example, in the development of biomimetic sonar systems for navigation, efforts should be made to avoid interference with the echolocation signals of bats and other species. This may involve employing frequency ranges that are outside the typical spectrum used by local wildlife or incorporating features that signal the presence of the technology to avoid disruption.

One key aspect of the wildlife-friendly sonar is the implementation of frequency modulation inspired by bat echolocation. Traditional sonar systems emit high-frequency pulses that can disturb marine mammals, leading to behavioral changes and potential harm. The biomimetic solution involves modulating the frequency range to avoid interference with the communication and navigation signals of marine life.

Adaptive Technology for Environmental Sensing

Another ethical consideration in this case study is the incorporation of adaptive technology that responds to environmental conditions. The sonar system can adjust its parameters based on factors such as water depth, temperature, and the presence of marine life. This adaptability ensures that the technology remains wildlife-friendly in diverse maritime environments.

The ethical considerations outlined in this chapter serve as a compass, guiding us toward a future where innovation harmonizes with the principles of wildlife conservation. The sonic navigators that grace our world, offer inspiration and insight into sustainable and responsible technological advancements.

The intersection of technology and nature calls for a conscientious approach that acknowledges the interconnectedness of all life forms on our planet. By embracing ethical considerations in the development and deployment of echolocation-inspired technologies, we not only honor the brilliance of bats but also contribute to a future where human ingenuity coexists harmoniously with the diverse tapestry of the natural world.

From the skies where bats gracefully dance through the night guided by the echoes of their calls, we plunge into the ocean's depths, where a new chapter unfolds beneath the waves. In the realm of marine marvels, we find ourselves captivated by the awe-inspiring tales of whales. These gentle giants navigate the ocean's expanses with a grace that belies their colossal size, utilizing intricate communication methods and remarkable abilities. Now, we embark on an aquatic journey to uncover the marvels of whale navigation, communication, and the profound interconnectedness that echoes through the watery realm.

Chapter 8:

Tubercles and Tribulations—Biomimetics of Humpback Whale Tubercles

Tucked within the intricate details of a humpback whale's flipper are tiny, peculiar bumps known as tubercles. These seemingly unassuming protuberances hold the key to an extraordinary aspect of nature's ingenuity, sparking a wave of exploration and inspiration in the realm of biomimetics. Let's delve into the tubercles and tribulations faced by humpback whales—from the challenges faced by these majestic mammals to the groundbreaking innovations inspired by their natural design. The mysteries that lie beneath the surface hold potential for remarkable biomimetic applications to emerge from these oceanic wonders.

Tubercle Anatomy and Hydrodynamics

Among the many enigmatic features that contribute to the humpback's prowess in the aquatic realm, the tubercles adorning its flipper are particularly captivating. These small, knob-like structures, arranged in a distinctive pattern along the leading edge of the whale's pectoral fins, hold the key to a remarkable combination of biomechanics and hydrodynamics.

The Purpose of Tubercles

As scientists scrutinize the anatomy of humpback whale tubercles, a clear purpose begins to emerge. Unlike the smooth and streamlined flippers of other whale species, the humpback's appendages showcase an ingenious difference. Each tubercle, resembling small bumps or knobs, interrupts the otherwise sleek surface of the flipper. The primary function of these tubercles is linked to the hydrodynamics of the whale's movement through water. From drag reduction to lift enhancement, the tubercles play a crucial role in shaping the hydrodynamic prowess of humpback whales.

Reducing Drag

Drag is the force that resists the motion of an object through a fluid, such as air or water (mrBrown, 2005). For aquatic creatures like the humpback whale, navigating through water involves overcoming the resistance posed by drag. The tubercles on the leading edge of the whale's flippers act as natural turbulators, disrupting the smooth flow of water around the flipper and mitigating the effects of drag.

Research indicates that the tubercles create tiny vortices along the flipper's surface. These vortices, or controlled swirls of water, serve to energize the boundary layer of fluid flowing over the flipper. By energizing the boundary layer, tubercles effectively delay the onset of turbulent flow and separation, reducing the overall drag experienced by the whale as it moves through the water (Natarajan et al., 2014).

This drag reduction mechanism is of particular importance for humpback whales during their long migrations. By minimizing drag, tubercles contribute to energy conservation, allowing these magnificent creatures to cover vast distances with remarkable efficiency. The biomimetic potential of this drag reduction strategy has not gone unnoticed by engineers seeking to improve the efficiency of various man-made structures moving through fluid mediums.

Enhancing Lift

While drag reduction is crucial for energy efficiency, the tubercles on humpback whale flippers also play a pivotal role in enhancing lift. Lift is the force that enables an object to rise or stay aloft, and for aquatic animals like whales, it is essential for buoyancy and effective swimming.

The tubercles contribute to lift through a phenomenon known as the stall delay. In traditional aerodynamics, a stall refers to the sudden loss of lift that occurs when the angle of attack of an airfoil becomes too steep. Humpback whale tubercles, however, alter the dynamics of stall onset, allowing the whale to maintain lift at higher angles of attack.

As the whale adjusts its flippers during different phases of swimming, the tubercles help to manage the flow of water over the flipper. This stall delay mechanism ensures that the flippers continue to generate lift even at steeper angles, enabling the humpback whale to execute agile maneuvers and maintain control in the water. This capability is especially crucial during activities such as feeding, where precision in movement is essential.

The biomimetic applications of this lift enhancement mechanism are far-reaching. Engineers and researchers have drawn inspiration from humpback whale tubercles to design more efficient hydrofoils, turbine blades, and other structures that interact with fluid mediums. By mimicking nature's design, these biomimetic innovations aim to improve the performance and sustainability of various human-made systems.

Biomimetic Insights in Fluid Dynamics

The intricate world of humpback whale tubercles opens a gateway to a realm where biology and technology unite. The biomimetic insights derived from the fluid dynamics of these magnificent mammals provide inspiration that transcends the depths of the ocean and reaches into the skies and across the landscapes of human innovation.

Aircraft Wing Design

The principles of fluid dynamics that govern the movement of humpback whales through water find unexpected applications in the design of aircraft wings. One of the key challenges in aviation is optimizing lift while minimizing drag, a quest that mirrors the aquatic tribulations faced by humpback whales.

Engineers and aerodynamicists have drawn inspiration from the tubercles on humpback whale flippers to enhance the performance of aircraft wings. By incorporating tubercle-like bumps along the leading edges of wings, researchers aim to achieve the same benefits observed in whale flippers—a reduction in drag and an enhancement in lift.

The biomimetic approach to aircraft wing design involves creating controlled turbulence along the wing surface, akin to the vortices generated by humpback whale tubercles. This turbulence delays the onset of stall, allowing aircraft to maintain lift at higher angles of attack. The result is a more efficient wing design that improves fuel efficiency, reduces emissions, and enhances the overall performance of aviation systems.

Wind Turbine Efficiency

In the pursuit of renewable energy, the wind energy sector has also turned to biomimicry inspired by humpback whale tubercles. Wind turbine blades, much like the flippers of marine mammals, face challenges related to drag and lift as they slice through the air.

A press release from the European Patent Office explores the integration of tubercle-inspired designs on wind turbine blades to enhance their efficiency. By disrupting the airflow and reducing drag, these biomimetic adaptations aim to improve the overall performance of wind turbines. The controlled vortices created by tubercle-like bumps on the blade surfaces contribute to a stall delay, ensuring that the turbine continues to harness energy even in turbulent wind conditions (*Turbine and Fans Inspired by Whales*, 2018).

The application of humpback whale tubercles in wind turbine design not only increases energy output but also addresses concerns related to noise and structural fatigue. As the renewable energy landscape evolves, biomimetic insights from the fluid dynamics of marine mammals pave the way for sustainable and ecologically sensitive innovations in wind power.

Advancements in Underwater Vehicles

Biomimetics inspired by humpback whale tubercles extend their influence beneath the ocean's surface, shaping the design and efficiency of underwater vehicles. Submarines and autonomous underwater vehicles (AUVs) encounter similar challenges to their airborne counterparts—the need for streamlined motion, reduced drag, and enhanced maneuverability.

By studying the hydrodynamic principles of humpback whale tubercles, engineers have developed biomimetic solutions for underwater vehicle design. The incorporation of tubercle-like structures on the surfaces of submarines and AUVs leads to a significant reduction in drag, allowing for smoother and more energy-efficient underwater travel.

Furthermore, the lift enhancement mechanism observed in humpback whale tubercles has inspired innovations in underwater vehicle control. Mimicking the stall delay effect, these vehicles can maintain stability and precision even at varying angles, expanding their capabilities for scientific exploration, oceanography, and defense applications.

The biomimetic journey inspired by humpback whale tubercles transcends the confines of oceanic exploration, reaching into the skies and beneath the waves. From aircraft wings that defy traditional aerodynamics to wind turbines harnessing the power of turbulence, and underwater vehicles gliding through the depths with newfound efficiency, the impact of fluid dynamics biomimicry is profound.

Ethical Considerations and Marine Conservation

The seamless integration of nature's brilliance into human technology demands a careful examination of our practices to ensure the protection and preservation of these magnificent marine creatures. This involves the ethical considerations surrounding humpback whales, the responsible use of tubercle-inspired designs, and sustainable energy generation that exemplifies the principles of marine conservation.

Protecting Humpback Whales

As we draw inspiration from nature's wonders, we must remain steadfast in our commitment to the preservation of marine life and the ecosystems that sustain it. The tubercles that have sparked innovations in fluid dynamics should serve as a reminder of the delicate balance between exploration and conservation. Humpback whales, like many marine species, face a myriad of threats ranging from climate change and habitat degradation to entanglement in fishing gear and collisions with vessels.

Conservation efforts must take precedence to ensure the well-being and survival of humpback whales. Strict regulations on whale watching, shipping lanes, and fishing practices are essential components of safeguarding these marine giants. Researchers and engineers engaged in biomimetic studies must collaborate with marine conservation organizations to promote practices that prioritize the welfare of humpback whales and their ecosystems.

Ethical Use of Tubercle-Inspired Designs

The translation of humpback whale tubercles into human technology carries with it a responsibility to use these designs ethically. While biomimetic applications offer tremendous potential for improving efficiency and sustainability, it is crucial to ensure that such innovations do not inadvertently harm the very ecosystems we seek to emulate.

In aviation, for example, the integration of tubercle-inspired designs in aircraft wings must adhere to rigorous safety and environmental standards. A thorough understanding of the potential consequences, both positive and negative, is essential in the development and deployment of these technologies.

Moreover, the use of biomimetic insights in wind turbine design must consider the ecological impact on bird and bat populations. The quest for cleaner energy should not come at the expense of other species, and thoughtful consideration of the ethical implications of our innovations is paramount.

Protecting humpback whales requires not only stringent conservation measures but also a cultural shift in how we view our relationship with the natural world. Ethical considerations must guide the development and implementation of tubercle-inspired designs, ensuring that our advancements contribute positively to the health and balance of marine ecosystems.

As we bid farewell to the awe-inspiring world of humpback whale tubercles and their biomimetic wonders, we embark on a new chapter of discovery, setting our sights on the mysterious realm inhabited by Cuvier's beaked whales. These deep-diving creatures, shrouded in an air of mystery, have the potential to unveil further secrets of the marvels that grace the vast expanses of our oceans.

Chapter 9:

Deep Divers—Biomimetics

of Cuvier's Beaked Whales

Among the marvels of mammals, there is perhaps no creature more enigmatic and awe-inspiring than the Cuvier's beaked whale. Named after the legendary French naturalist Georges Cuvier, these whales are the embodiment of resilience and adaptation in the harsh, mysterious depths of the ocean. With sleek, torpedo-like bodies and distinctive beaks, they navigate the abyss with unparalleled grace and efficiency. What makes them stand out, however, is their remarkable ability to plunge into the ocean's depths and endure prolonged dives that last for more than two hours.

From their physiological marvels that enable extended breath-holding to their navigational prowess in the oceanic darkness, these deep divers have captivated scientists and enthusiasts alike. Let's uncover the secrets behind their deep-sea survival, shedding light on the incredible biomimetic potential that these champions of the abyss hold for technological inspiration and understanding the mysteries of the deep.

The Enigmatic Cuvier's Beaked Whales

Cuvier's beaked whales are the undisputed kings of deep-sea diving. What sets them apart from other marine mammals is their astounding ability to plunge into the ocean's depths with unparalleled efficiency. These whales have been recorded diving to depths exceeding 9,800 feet (3,000 meters), navigating the abyss with a grace that captivates marine biologists and enthusiasts alike.

Deep Diving Abilities and the Marvel of Collapsing Ribs and Lungs

One of the key features that enable these deep dives is the unique structure of their respiratory system. Unlike most mammals, Cuvier's beaked whales possess the remarkable ability to collapse their ribs and lungs during deep dives. This physiological marvel allows them to minimize buoyancy and decrease the risk of decompression sickness, a condition caused by rapid changes in pressure. As they descend into the inky blackness of the ocean, their bodies undergo a transformative process, adapting to the crushing pressures of the deep.

The collapsing of ribs and lungs serve a dual purpose—not only do they enhance the whale's diving capabilities, but they also aid in reducing the amount of nitrogen absorbed into the whale's bloodstream. This ability is crucial for avoiding the bends, a condition where nitrogen bubbles form in the blood and tissues due to rapid ascent. Cuvier's beaked whales have mastered the art of controlled submersion, allowing them to explore the depths with unparalleled precision.

Oxygen Utilization and the Ability to Store Oxygen in Muscles

Surviving in the profound depths requires efficient oxygen utilization, and Cuvier's beaked whales have unique features needed to excel in this aspect. These marine marvels exhibit exceptional breath-holding capabilities, with some individuals recorded holding their breath for more than two hours during a single dive.

One of the secrets behind their extended breath-holding is their ability to store oxygen not only in their blood but also in their muscles. While most mammals primarily rely on oxygen stored in the bloodstream, Cuvier's beaked whales take it a step further. Their muscles, particularly the myoglobin-rich muscles, act as oxygen reservoirs, allowing them to sustain prolonged dives without succumbing to hypoxia.

Myoglobin is a protein found in muscles that binds with oxygen, facilitating its storage. The muscles of Cuvier's beaked whales boast high concentrations of myoglobin, enabling them to extract and utilize oxygen more efficiently. This innate ability enhances their endurance, making them true masters of the deep-sea plunge.

Pressure Tolerance and the Astonishing Depths of Exploration

Surviving in the crushing depths of the ocean requires an exceptional tolerance for pressure, and Cuvier's beaked whales can withstand conditions that would be lethal for most mammals, pushing the limits of what was once thought impossible for marine mammals.

Their bodies are designed to cope with the extreme pressure encountered in the abyss. The collapsible rib cage and lung adaptation not only aids in minimizing buoyancy but also protects them from the effects of pressure changes. Additionally, their physiology allows for efficient equalization of pressure, preventing barotrauma, a condition caused by pressure differences between the inside and outside of the body.

As they descend into the depths, Cuvier's beaked whales navigate a world of darkness, utilizing their echolocation abilities to locate prey. The pressure tolerance exhibited by these deep divers is a testament to the marvels of these animals, showcasing how nature has crafted a creature perfectly suited for the challenges of the deep.

Diving Behavior and Biomimetic Applications—Unveiling the Secrets of Cuvier's Beaked Whales

Cuvier's beaked whales engage in a dance of survival, showcasing unparalleled diving behavior that has captivated scientists and inspired biomimetic innovations. Their hunting strategies, record-breaking dives, and potential applications for projects unlock a world of possibilities hidden within the realm of these deep-sea champions.

Hunting in the Depths—Group Dynamics Unveiled

Cuvier's beaked whales are not solitary deep divers; instead, they navigate the ocean depths in cohesive groups, displaying intricate social structures and cooperative hunting behaviors. These marine marvels often engage in synchronized dives, creating a choreography of motion as they descend into the abyss in search of prey.

Group hunting serves several purposes for Cuvier's beaked whales. It allows them to cover larger areas in search of elusive deep-sea squid, their primary prey. As they dive, these whales use echolocation to detect prey in the impenetrable darkness, emitting sonar clicks that bounce off objects in the water. The group dynamic enhances the effectiveness of this echolocation, creating a collaborative effort to locate and capture prey.

The complex group dynamics observed in Cuvier's beaked whales have inspired researchers to explore biomimetic applications for underwater robotics and swarm intelligence. Mimicking the synchronized movements and communication observed in these deep-sea hunters, engineers are developing autonomous underwater vehicles that can work together seamlessly, opening up new possibilities for ocean exploration and resource monitoring.

Record-Breaking Dives—Nature's Inspiration for Extreme Exploration

Cuvier's beaked whales hold the record for some of the deepest and longest dives ever recorded among marine mammals. Scientists and engineers alike are drawn to the challenge of understanding and emulating these record-breaking dives. By unraveling the physiological adaptations that enable these remarkable dives, researchers aim to develop technologies that can withstand the extreme conditions of the deep ocean.

One area of biomimetic research focuses on the development of materials and structures that can withstand high pressures experienced in the deep-sea environment. The collapsible rib cage and lung, which allows the whales to minimize buoyancy and navigate the depths efficiently, has inspired advancements in materials science for the design of deep-sea submersibles and underwater habitats (Blasiak et al.,2022).

Innovations in Autonomous Underwater Vehicles—Navigating the Abyss

The exploration of the deep sea poses numerous challenges, from extreme pressure conditions to the vastness of uncharted territories. Inspired by the diving prowess of Cuvier's beaked whales, researchers are harnessing biomimetic principles to revolutionize the field of autonomous underwater vehicles (AUVs).

One key aspect of biomimicry involves emulating the hydrodynamics of the whales' streamlined bodies. The sleek, torpedo-like shape of Cuvier's beaked whales minimizes drag, allowing them to move efficiently through the water. Engineers are incorporating these streamlined designs into the development of AUVs, enhancing their maneuverability and range for deep-sea exploration.

Additionally, the whales' ability to collapse their ribs and lungs during dives has implications for the design of AUVs that need to navigate varying depths. By incorporating adaptable structures that mimic this collapsing mechanism, researchers aim to create underwater vehicles capable of adjusting their buoyancy in response to changing conditions, offering a level of flexibility crucial for deep-sea exploration.

Furthermore, the advanced echolocation capabilities exhibited by Cuvier's beaked whales have spurred developments in underwater sensing technologies. Engineers are working on refining sonar systems inspired by the whales' natural echolocation, enhancing the ability of AUVs to detect and navigate through complex underwater environments.

From group dynamics influencing swarm robotics to record-breaking dives inspiring advancements in materials science and autonomous vehicles, the secrets hidden within the deep-sea behaviors of these whales are shaping the future of ocean exploration.

Ethical Considerations and Ocean Conservation—The Guardianship of Cuvier's Beaked Whales

These deep-sea marvels, while inspiring technological advancements, also demand our commitment to protecting their habitats and ensuring the sustainability of our exploratory endeavors. As we delve into the biomimetics of Cuvier's beaked whales, it is imperative to navigate the waters of ethical considerations and emphasize the crucial role they play in the broader context of ocean conservation.

Protecting Deep-Sea Ecosystems—Balancing Curiosity With Conservation

The exploration of the deep sea, driven by our insatiable curiosity and the pursuit of knowledge, must be approached with a deep sense of responsibility. Cuvier's beaked whales are intricately connected to the health of deep-sea ecosystems. Understanding and emulating their remarkable abilities should not come at the cost of jeopardizing the delicate balance of these underwater environments.

Conservation efforts must extend beyond the charismatic whales themselves to encompass the broader ecosystems they inhabit. Deep-sea habitats are often fragile and slow to recover from disturbances. The deployment of autonomous underwater vehicles (AUVs) and other exploration technologies must be done with meticulous care to avoid unintentional damage to the delicate ecosystems and the myriad of species that call the depths home.

Initiatives such as marine protected areas (MPAs) and international collaborations to regulate deep-sea exploration are crucial steps toward safeguarding these ecosystems. By setting boundaries on human activities in designated areas, we can mitigate the potential ecological impact of our technological advancements and ensure the longevity of the deep-sea wonders that continue to inspire our scientific pursuits.

Research Initiatives—Balancing Scientific Progress With Ethical Responsibility

Understanding the biology and behavior of Cuvier's beaked whales involves intensive research efforts, including tagging, tracking, and sometimes, the unfortunate discovery of stranded individuals. Researchers studying Cuvier's beaked whales are increasingly adopting non-invasive methods, such as satellite tagging and passive acoustic monitoring, to gather information without directly impacting the animals. These approaches respect the whales' natural behaviors and contribute to a growing body of knowledge that informs conservation strategies.

Doctors Treating Collapsed Lungs— Bridging Medicine and Conservation

The unique ability of Cuvier's beaked whales to collapse their ribs and lungs during deep dives has intrigued medical professionals seeking innovative solutions for treating collapsed lungs in humans. However, this intersection of biomimetics and medicine raises ethical considerations that require careful navigation.

Researchers and doctors working on medical applications inspired by the whales' abilities must prioritize ethical research practices and acknowledge the potential consequences of their work on the welfare of these marine mammals. Collaboration between marine biologists, engineers, and medical professionals becomes crucial to strike a balance between advancing human healthcare and preserving the integrity of deep-sea ecosystems.

By striking a balance between scientific curiosity, technological progress, and conservation efforts, we can ensure that our exploration of the deep sea leaves a positive and lasting impact on the oceans and the extraordinary creatures that inhabit them. Cuvier's beaked whales, guardians of the abyss, beckon us to explore with reverence and responsibility, nurturing a harmonious relationship between humanity and the wonders of the deep.

Now, our marine journey continues, weaving through the intricate tapestry of oceanic marvels, as we shift our focus from the enigmatic depths with Cuvier's beaked whales to the sleek, sunlit surfaces where the long-finned pilot whales gracefully navigate. In the boundless expanse of the sea, these creatures introduce us to a fascinating realm where innovation takes the form of self-cleaning skin—a feature that not only defines their biological resilience but also holds profound implications for biomimetic exploration.

Chapter 10:

Masters of Self-Cleaning

Skin—Biomimetics of

Long-Finned Pilot Whales

In the vast expanse of the ocean, long-finned pilot whales reign supreme as the masters of self-cleaning skin. These enigmatic creatures have captivated researchers and enthusiasts alike with their unique biomimetic qualities, showcasing a remarkable ability to keep their skin pristine in the challenging underwater environment.

These whales' ability to uphold the pristine condition of their skin is a testament to the wonders of adaptation and serves as an inspiration for understanding nature's ingenious solutions to the challenges posed by the marine world. Let's unravel the secrets behind the pilot whales' exceptional skincare routine and delve into the intricacies of their self-maintenance techniques. Let the journey into the realm of these self-cleaning masters unfold, revealing the secrets hidden beneath the waves.

Unveiling the Skin Microbiome and Functionality in Long-Finned Pilot Whales

The skin of long-finned pilot whales serves as a dynamic ecosystem, hosting a myriad of beneficial microorganisms that contribute to their overall health. These microscopic inhabitants form a complex community, establishing a symbiotic relationship with the whale. Among these inhabitants are bacteria that act as natural defenders, protecting the whale from harmful pathogens.

Research has identified specific strains of bacteria on the skin of long-finned pilot whales that exhibit antimicrobial properties. These friendly bacteria create a biological shield, preventing the colonization of harmful microorganisms and pathogens (AskNature Team, n.d.). The delicate balance within the skin microbiome highlights the intricate relationship between these microorganisms and the host, showcasing a remarkable coexistence.

Antimicrobial Properties

Pilot whales boast a unique set of antimicrobial properties embedded within their skin. These properties are attributed to the secretion of specialized compounds that inhibit the growth of harmful bacteria and fungi. The whales produce natural antibiotics, effectively warding off potential threats to their skin health.

Understanding the antimicrobial arsenal of pilot whales offers valuable insights into biomimetics, inspiring the development of innovative solutions in human skincare and medical fields. The study of these natural defenses not only enhances our knowledge of marine biology but also opens doors to potential breakthroughs in combating antibiotic-resistant pathogens.

Self-Cleaning Behavior and Nanoridges Filled With Gel

One of the most intriguing aspects of the pilot whales' self-cleaning prowess lies in their behavior and the unique nanostructures present on their skin. These marine mammals exhibit a fascinating self-cleaning behavior, regularly engaging in rubbing against surfaces such as rocks and other whales. This behavior serves as a proactive approach to remove debris and parasites from their skin.

The nanoridges found on the skin of pilot whales play a crucial role in their self-cleaning mechanism. These microscopic ridges create a textured surface that minimizes friction, allowing water to easily glide over the skin. Moreover, these nanoridges are filled with a specialized gel-like substance that enhances the self-cleaning process. As the whales move through the water, the gel assists in dislodging and repelling particles, leaving the skin virtually spotless.

Biomimetics Inspired by Long-Finned Pilot Whales

The long-finned pilot whales' skin microbiome and functionality provide a wealth of inspiration for biomimetics. Scientists and engineers are exploring the integration of beneficial microorganisms, antimicrobial compounds, and nanoridge structures in the development of advanced materials and products.

The utilization of pilot whales' self-cleaning strategies in biomimetic design holds tremendous potential across various industries. From self-cleaning surfaces in architecture to innovative medical applications, the lessons learned from these marine marvels pave the way for sustainable and efficient solutions inspired by nature. The harmonious relationship between beneficial microorganisms, antimicrobial properties, and the ingenious nanoridges filled with gel also reveals a world of possibilities for biomimetic applications.

Communication Through Skin

The skin of pilot whales serves as a canvas for communication within their pods. These cetaceans engage in a form of non-vocal communication through various skin-related behaviors. Among these is the intriguing phenomenon of "spy hopping," where whales lift their heads above the water's surface to observe their surroundings. During this process, the distinctive patterns and markings on their skin become visible, allowing for visual recognition and potential signaling to other members of the pod.

Furthermore, pilot whales possess a remarkable ability to modulate the coloration of their skin, a feature often associated with emotional states and social interactions. Changes in skin color can convey information about the whale's mood, intentions, and responses to external stimuli. The intricacies of skin communication in these marine mammals underscore the importance of skincare not only as a physical necessity but also as a means of expressing and interpreting social cues.

Social Bonding and Mutual Grooming

Skincare extends beyond the realm of hygiene for long-finned pilot whales; it's also a powerful tool for social bonding. Mutual grooming, a behavior observed extensively within whale pods, involves individuals using their teeth to gently scrape the skin of their companions. This ritualistic grooming serves multiple purposes, including the removal of dead skin and parasites, but it transcends mere hygiene.

Mutual grooming fosters a sense of intimacy and trust within the pod. It is a tactile expression of social bonds, emphasizing the interconnectedness of individuals within the community. Through this physical interaction, long-finned pilot whales strengthen their social fabric, reinforcing alliances and kinship ties. The act of mutual grooming goes beyond the functional aspects of skincare; it becomes a language of connection and belonging among these majestic marine beings.

Role in Group Dynamics

Skincare plays a pivotal role in shaping the intricate dynamics of pilot whale pods. The social structure of these communities is characterized by strong familial bonds, with matriarchs often leading the way. Within these family units, the act of caring for one another's skin becomes a manifestation of group cohesion and cooperation.

Observations of long-finned pilot whale pods reveal that individuals, irrespective of age or hierarchical position, actively participate in the grooming process. The egalitarian nature of mutual grooming contributes to a sense of collective responsibility for the well-being of the entire pod. This cooperative behavior not only ensures the health of individual members but also reinforces the resilience and unity of the entire social group.

Social Implications for Human Society

The social implications of skincare in long-finned pilot whales prompt reflection on the parallels that can be drawn with human society. While our modes of communication differ, the importance of physical touch, grooming, and shared care rituals resonate across species. The mutual grooming observed in these marine mammals inspires contemplation on the significance of touch in human relationships, fostering emotional bonds and a sense of belonging.

Moreover, the collaborative nature of skincare within pilot whale pods serves as a poignant reminder of the strength found in community support. In an era where individualism often takes precedence, the communal aspects of skincare among these marine beings offer valuable insights into the potential benefits of collective well-being and shared responsibilities in human societies.

The exploration of the social implications of skincare in pilot whales unveils a rich tapestry of communication, bonding, and group dynamics. Beyond the physical attributes of self-cleaning skin, the rituals of mutual grooming become a language of connection, forging strong social bonds within their pods.

Bridging the Gap—Human Applications and Conservation Inspired by Long-Finned Pilot Whales

The self-cleaning prowess of pilot whales not only captivates our imaginations but also serves as a wellspring of inspiration for human innovation and conservation efforts.

Biomedical Discoveries

The unique skin features of long-finned pilot whales have spurred groundbreaking discoveries in the field of biomedical engineering. Inspired by the whales' self-cleaning nanoridges and gel-filled structures, scientists have developed a revolutionary metal mesh with an array of holes. This mesh, when exposed to seawater, exudes a biosafe chemical that undergoes a remarkable transformation. It thickens into a viscous gel, forming a protective layer over ship hulls (Castillote, 2020).

This biomimetic approach mimics the long-finned pilot whales' self-cleaning behavior, preventing the accumulation of marine organisms on ship surfaces. The gel-coated hulls, inspired by nature's ingenious design, significantly reduce drag and biofouling, leading to enhanced fuel efficiency and reduced environmental impact in maritime transportation. This innovative solution not only benefits the shipping industry but also exemplifies how lessons from marine life can be harnessed to address real-world challenges.

Conservation Challenges and Ethical Human-Animal Interaction

While biomimetics offers promising solutions, the conservation of pilot whales also presents a set of challenges. Human activities, such as noise pollution, entanglement in fishing gear, and habitat degradation, pose threats to these marine marvels. As we draw inspiration from their self-cleaning skin, it becomes imperative to address the broader conservation context and promote ethical human-animal interaction.

Conservation efforts should prioritize the mitigation of anthropogenic impacts on pilot whales' habitats. Initiatives to reduce maritime noise pollution, implement sustainable fishing practices, and establish marine protected areas contribute to the overall well-being of these cetaceans. Additionally, fostering awareness and promoting responsible whale-watching practices are crucial steps in ensuring ethical human-animal interaction, allowing these creatures to thrive in their natural environment.

Holistic Conservation Approach

To truly honor the marvels of long-finned pilot whales and their self-cleaning skin, a holistic conservation approach is essential. This entails not only harnessing biomimetics for human applications but also actively participating in the protection and restoration of marine ecosystems. The conservation of these marine marvels requires collaborative efforts at local, regional, and global levels.

Educational programs, research initiatives, and community engagement play pivotal roles in fostering a deeper understanding of pilot whales and the ecosystems they inhabit. By integrating biomimicry into marine conservation practices, we not only benefit from innovative solutions but also contribute to the preservation of biodiversity and the delicate balance of our oceans. As we navigate the intricate interplay between human ingenuity and the natural world, long-finned pilot whales beckon us to embark on a journey of discovery and preservation.

From the depths of the ocean, we now surface to embark on a journey that unveils the intelligence and sonar abilities of dolphins. Let's transition to the mysteries of the ocean's acoustic virtuosos, exploring the intricacies of their cognitive brilliance and the fascinating echolocation skills that have earned them the title of marine marvel.

Chapter 11:

Echolocation Experts—

Biomimetics of Dolphins'

Sonar

Dolphins, the virtuosos of underwater sonar, captivate us with their remarkable ability to harness echolocation for a myriad of purposes. They employ a sophisticated echolocation system that surpasses human comprehension. These marine marvels have perfected the art of navigating through the vast oceans, communicating with one another, and executing precision hunts—all guided by the intricate echoes bouncing through the underwater realm. Let's delve into the fascinating world of these echolocation experts, exploring the biomimetics of dolphins' sonar systems, a symphony that echoes the brilliance of nature's engineering.

Cephalic Air Sacs and Sound Production

At the heart of a dolphin's echolocation prowess lies its advanced sonar anatomy. Dolphins possess specialized structures that enable them to produce and receive sound with unparalleled precision. One key component is the complex system of air sacs located in their heads, known as cephalic air sacs. These air sacs play a crucial role in the production of sound waves.

When a dolphin initiates echolocation, it releases a burst of air through its blowhole, directing it into the nasal passages. The cephalic air sacs act as a sophisticated set of acoustic lenses, focusing and modulating the emitted clicks. These clicks, emitted at frequencies beyond the range of human hearing, travel through the water in search of objects and obstacles.

The ability to manipulate and control the characteristics of these sound waves allows dolphins to create a detailed and comprehensive sonar picture of their surroundings. It's a testament to the marvel of their anatomical makeup, enabling them to navigate the complex underwater environment with unparalleled precision.

Receiving and Processing Echoes

Echolocation is a two-way street for dolphins—not only do they emit clicks to probe their surroundings, but they also possess an extraordinary ability to receive and process the resulting echoes. The process begins with the emission of a click, which travels through the water and interacts with objects in its path. Upon encountering an obstacle, the sound waves bounce back as echoes, providing crucial information to the dolphin.

Dolphins have highly sensitive auditory systems, with specialized structures in their inner ears designed to receive and interpret these echoes. The intricate cochlea and associated neural pathways enable them to discriminate between subtle differences in echo timing, intensity, and frequency.

This keen ability allows dolphins to distinguish between objects with astonishing precision. Picture a fish concealed in the sandy ocean floor—a challenge for many other marine species. For a dolphin, however, the echolocation system enables them to perceive the fish's location and even differentiate between the sand particles and the concealed prey. It's a testament to their ability to navigate and hunt efficiently in environments where visibility is limited.

Underwater Identification

The echolocation expertise of dolphins extends beyond mere obstacle detection; it encompasses a remarkable capability to identify and discriminate between objects based on their internal composition. One notable example involves dolphins being able to distinguish between the contents inside a sealed barrel submerged underwater.

When presented with such a scenario, dolphins emit clicks directed toward the sealed barrel. The echoes returning from the object carry information about its internal structure. Dolphins, through their highly intelligent auditory processing, can differentiate between various materials and even identify the nature of the contents inside the sealed container.

This level of underwater identification goes beyond mere echolocation; it speaks to the intricate neural processing that occurs within a dolphin's brain. The ability to discern the subtle acoustic signatures of different materials showcases the incredible cognitive and sensory capacities of these marine mammals.

Biomimetic Applications in Underwater Acoustics

The mesmerizing echolocation abilities of dolphins have not only captured the fascination of marine enthusiasts but have also inspired a wave of biomimetic applications in the realm of underwater acoustics. Let's explore the innovative ways in which human researchers have drawn inspiration from dolphins' sonar systems to develop cutting-edge technologies with diverse applications, ranging from locating buried infrastructure to advancements in medical imaging and marine research.

Human Applications

One of the remarkable biomimetic applications derived from dolphins' echolocation expertise lies in the realm of locating buried cables and pipelines beneath the seabed. Human researchers have been inspired by the precision with which dolphins navigate and identify objects underwater, leading to the development of devices that harness similar principles for practical applications.

These biomimetic devices utilize acoustic signals to penetrate the seabed and detect the presence of buried cables and pipelines. By emitting sound waves and analyzing the returning echoes, these technologies can create detailed maps of the underwater landscape, identifying the exact location and depth of buried infrastructure. This has proven invaluable for industries involved in offshore construction, maintenance, and repair, as it allows for the precise planning and execution of activities without the need for extensive excavation.

The biomimetic approach not only enhances efficiency but also minimizes environmental impact, showcasing how insights from nature can be harnessed to address human challenges sustainably.

Medical Imaging Technologies

Dolphins' echolocation capabilities have inspired breakthroughs in the field of medical imaging, offering a new perspective on non-invasive diagnostic techniques. Research has explored the biomimetic application of underwater acoustics to develop innovative imaging technologies that draw parallels with the precision of dolphins' sonar systems (*Dolphin-Inspired Compact Sonar*, 2023).

One notable example is the development of acoustic imaging devices that utilize sound waves to create detailed images of internal structures within the human body. Similar to how dolphins discern objects underwater, these devices emit sound waves that penetrate tissues and organs, generating echoes that are then processed into high-resolution images.

This biomimetic approach holds promise for enhancing medical diagnostics, enabling healthcare professionals to visualize internal structures with greater clarity and accuracy. The non-invasive nature of these techniques reduces patient discomfort and provides a valuable tool for early detection and monitoring of various medical conditions.

Advancements in Marine Research

The biomimetic applications of dolphins' echolocation extend beyond practical utilities to contribute significantly to the field of marine research. Researchers have leveraged insights from dolphin sonar systems to enhance our understanding of underwater environments and the diverse marine life inhabiting them.

Underwater drones equipped with biomimetic sonar systems have been deployed to explore the depths of the ocean with unprecedented precision. These drones emulate the echolocation abilities of dolphins, allowing researchers to map the topography of the ocean floor, identify marine species, and study underwater ecosystems in ways previously unattainable.

The biomimetic advancements in marine research have opened new avenues for exploration and conservation. By applying principles inspired by dolphins, scientists can conduct non-intrusive studies of marine life and ecosystems, contributing to our knowledge of the oceans and informing conservation efforts.

The biomimetic applications of dolphins' sonar systems in underwater acoustics have ushered in a new era of innovation and discovery. As researchers continue to draw inspiration from these marine marvels, the potential for biomimetic solutions to address diverse challenges and push the boundaries of human knowledge becomes increasingly evident.

Ethical Considerations and Marine Conservation

Dolphins play a vital role in maintaining the balance of marine ecosystems. Their sophisticated sonar abilities not only aid in navigation and hunting but also contribute to the overall health of the oceans. As we marvel at their capabilities, it is crucial to recognize the significance of dolphin conservation for the well-being of our planet.

Dolphins face various threats, including habitat degradation, pollution, entanglement in fishing gear, and the disturbance caused by human activities such as shipping and recreational boating. Conservation efforts are essential to protect these intelligent marine mammals and the ecosystems they inhabit. Preserving their natural habitats, implementing sustainable fishing practices, and minimizing human-induced disturbances are key components of ensuring the continued existence of these marvels of the ocean.

Ethical Use of Sonar-Inspired Technologies

While the biomimetic applications of dolphins' sonar abilities hold great promise for various industries, it is crucial to approach these technologies with ethical considerations in mind. The use of sonar-inspired technologies must be guided by principles that prioritize the well-being of marine life and the conservation of natural ecosystems.

One ethical concern involves the potential impact of active sonar on marine mammals, including dolphins. Active sonar systems emit intense pulses of sound, and there is evidence to suggest that exposure to high-intensity sonar can disrupt the behavior and communication of marine mammals, leading to issues such as strandings and changes in feeding habits.

To address these concerns, ethical guidelines must be established for the deployment of sonar technologies in marine environments. These guidelines may include limitations on sound intensity, seasonal restrictions to avoid sensitive periods in marine life cycles, and the incorporation of real-time monitoring to detect and mitigate potential impacts.

From the playful world of dolphins, we shift our focus to the intricate tapestry of fur that makes sea otters one of the most captivating creatures in the animal kingdom. Let's now explore the wonders of sea otter fur, a testament to nature's ingenuity and the diversity of marvels that grace our planet's ecosystems.

Chapter 12:

Fur for All Seasons—

Biomimetics of Sea Otter

Fur

In the vast expanse of cold and unforgiving waters, where the whims of nature dictate the rules of survival, aquatic mammals have ingenious mechanisms to combat the relentless chill. Among these resilient inhabitants, the sea otter emerges as a captivating protagonist with a unique tale of intelligence—one that revolves around the extraordinary prowess of its fur.

Let's unfold a journey through the seasons, exploring how the sea otter's fur becomes a versatile tool for survival in changing environments. From the icy waters of winter to the milder currents of spring, the sea otter's fur adapts seamlessly, showcasing the true essence of biomimetics. The incredible insulating properties of sea otter fur stand as a testament to nature's ingenuity, offering a glimpse into the endless possibilities that arise when we observe, learn, and draw inspiration from the marvels of the animal kingdom.

Fur Anatomy and Structure

While many of its marine counterparts rely on a thick layer of blubber beneath the skin to insulate against the biting cold, the sea otter has a different path, one that involves the mastery of fur as its primary means of staying warm.

Sea otters boast one of the densest fur coverings in the animal kingdom. Their fur is comprised of two layers: the outer guard hairs and the inner insulating underfur. The outer guard hairs are longer and coarser and provide a waterproof barrier, preventing the inner fur from becoming saturated. This feature is crucial for sea otters, as they spend a significant amount of time in water.

The underfur, on the other hand, is exceptionally dense, with about 600,000 to 1,000,000 hair follicles per square inch. This incredible density creates an insulating layer, trapping a layer of air close to the otter's skin. Air is an excellent insulator, and by keeping it trapped in their fur, sea otters can maintain a stable body temperature even in chilly waters.

The unique structure of sea otter fur also includes specialized hairs called vibrissae, which are highly sensitive and help the otters detect prey underwater. These vibrissae play a crucial role in the otters' hunting success, allowing them to navigate and locate food with remarkable precision.

Thermal Insulation Abilities

The ability of sea otters to maintain their body temperature is particularly crucial during colder seasons when the surrounding water temperatures drop. The fur provides an effective barrier against the cold, allowing otters to spend prolonged periods in the water without succumbing to hypothermia. Its insulating properties are nothing short of extraordinary, allowing these marine mammals to maintain a comfortable body temperature in waters that would otherwise be unbearably cold.

Energy-Efficient Building Materials

The dense and insulating nature of sea otter fur has inspired researchers to explore its applications in energy-efficient building materials. Mimicking the structure of otter fur, architects and engineers are developing materials that can provide effective thermal insulation for buildings, reducing the need for extensive heating or cooling systems.

One approach involves creating synthetic materials with a microstructure similar to sea otter fur, utilizing microscopic air pockets to enhance insulation. These materials can be incorporated into walls, roofs, and other architectural elements to regulate indoor temperatures more efficiently. By drawing inspiration from nature's design, biomimetic building materials aim to decrease energy consumption in structures, contributing to sustainability and environmental conservation.

Cold-Weather Clothing Innovations

Sea otters excel in maintaining their body temperature in frigid waters, thanks to their fur's insulation properties. This has led to biomimetic insights in the realm of cold-weather clothing. Researchers and designers are exploring ways to integrate otter fur-inspired technologies into winter garments, creating more effective and lightweight insulation.

Microfiber technologies that mimic the structure of sea otter fur are being developed to enhance the warmth and comfort of cold-weather clothing. These synthetic fibers trap air close to the body, providing superior insulation without the bulkiness associated with traditional materials. The biomimetic approach not only improves the performance of cold-weather attire but also promotes sustainability by reducing the reliance on heavy, non-biodegradable insulation materials.

Advancements in Sustainable Textiles

The sustainable textile industry has embraced biomimicry as a means to reduce environmental impact and promote eco-friendly practices. Sea otter fur has inspired the development of sustainable textiles that prioritize both functionality and environmental responsibility.

Innovations in fabric production draw inspiration from the otter's fur to create water-resistant and insulating textiles without relying on harmful chemicals or environmentally detrimental processes. These textiles find applications in outdoor gear, sportswear, and everyday clothing. The biomimetic approach in sustainable textiles aligns with the growing demand for eco-conscious products, offering consumers alternatives that prioritize performance without compromising the planet's well-being.

Challenges and Future Directions

While biomimetic innovations inspired by sea otter fur hold great promise, challenges remain in translating these insights into scalable and cost-effective solutions. The intricate structure of otter fur presents engineering challenges that require meticulous research and development. Additionally, ensuring that biomimetic technologies align with sustainability goals and ethical considerations is essential for their long-term viability.

Future directions in biomimetic insulation technology involve interdisciplinary collaborations between biologists, engineers, and materials scientists. By deepening our understanding of the principles governing sea otter fur, researchers can refine biomimetic approaches and explore new applications in diverse fields.

Conservation Concerns

The exploration of sea otter fur and its biomimetic applications in various industries raises important ethical considerations and emphasizes the critical need for wildlife conservation efforts. While sea otters showcase an extraordinary adaptation in their fur, their populations face numerous conservation concerns. Historically, sea otters were hunted extensively for their fur, leading to a significant decline in their numbers. Conservation efforts have helped in some regions, but sea otters are still listed as endangered or threatened in various areas.

The oil industry poses a significant threat to sea otters, as oil spills can compromise the waterproofing abilities of their fur. When otters come into contact with oil, their fur loses its insulating properties, making them susceptible to hypothermia. Coastal development, pollution, and entanglement in fishing gear also contribute to the challenges faced by sea otter populations.

Sea otters, as keystone species, exemplify the far-reaching impacts of their existence on coastal ecosystems. By controlling sea urchin populations through their foraging activities, otters indirectly support the health of kelp forests. These underwater ecosystems, in turn, provide habitat and sustenance for numerous marine species. The ripple effects of sea otter activities demonstrate how the well-being of one species can reverberate throughout the interconnected tapestry of life.

Efforts to protect and conserve sea otters involve strict regulations on hunting, habitat restoration, and education on the importance of these keystone species in coastal ecosystems. Conservationists and researchers continue to monitor sea otter populations, striving to ensure the long-term survival of these marvelous mammals and their exceptional fur.

Ethical Use of Fur-Inspired Designs

As biomimetic innovations inspired by sea otter fur gain traction, ethical considerations become paramount in ensuring that the principles of conservation and animal welfare are upheld. The use of fur-inspired designs in various industries, such as fashion and technology, requires scrutiny to avoid perpetuating harm to wildlife.

Ethical guidelines should be established to determine the responsible sourcing of materials and the humane treatment of animals. Synthetic alternatives that mimic the properties of sea otter fur without directly utilizing animal products offer a more ethical approach. Additionally, supporting and promoting sustainable practices within industries can mitigate the environmental impact associated with production processes.

Transparent labeling and certification systems can help consumers make informed choices, ensuring that products align with ethical standards. Industry collaborations with conservation organizations can further emphasize the commitment to ethical considerations, fostering a culture of responsible innovation.

As we continue our odyssey through this extraordinary realm, a profound sense of wonder propels us forward. Let's embark into the realm of limbs, vision, skin, and taste, discovering the remarkable similarities and ingenious abilities that connect us with the vast tapestry of existence. We'll discover a breathtaking convergence of design principles, where the threads of similarity weave through the fabric of existence. These fundamental aspects connect us with the broader web of life, inviting us to marvel at the shared ingenuity that defines the living world.

Chapter 13:

Nature's Blueprint—

Unveiling Marvels in

Biomimetics: Limbs,

Vision, Skin, and Taste

In the grand tapestry of life, mammals stand as extraordinary marvels, each species sculpted by the relentless forces of advancement into unique and wondrous beings. Among the many enchanting features that define mammals, their limbs, vision, skin, and sense of taste emerge as awe-inspiring masterpieces of nature's ingenuity. Let's explore how science draws inspiration from each of these aspects of the animal kingdom to forge innovations that echo the brilliance of natural design.

Limbs of Wonder—Biomimetics of Human and Animal Limbs

From the wings of a bat that echo the grace of a nocturnal ballet to the powerful legs of a kangaroo, designed for boundless leaps across the Australian outback, the diversity of mammalian limbs unfolds like a mesmerizing dance. In our exploration of biomimetics, we uncover the secrets of nature's engineering prowess, delving into how scientists and engineers draw inspiration from these limbs to create prosthetics, robotics, and exoskeletons that redefine the boundaries of human potential.

Locomotion and Mobility

Mammalian limbs are the engines of motion, propelling creatures through landscapes as diverse as the Himalayan mountains and the Amazon rainforest. Nature presents us with the balletic leaps of gazelles as they evade predators on the African plains or the synchronized swimming of dolphins gracefully slicing through the ocean currents. Whether it's the acrobatic antics of tree-dwelling lemurs or the marathon runs of mice, each species reveals a chapter in the remarkable story of mammalian mobility.

Unique Appendages

Survival is a dynamic dance, and mammalian limbs have donned an array of costumes to thrive in their respective roles. For example, the elongated fingers of a bat, perfect for grasping insects mid-flight, or the opposable thumbs of primates, enabling intricate tool manipulation. The retractable claws of big cats, a predatory innovation, or the hoofed appendages of ungulates, adapted for efficient travel across varied terrains. Unique appendages not only reflect the diversity of mammalian life but also the intricate ways limbs are tailored to the challenges presented by their ecological niches.

Biomimetic Applications

As humanity pushes the boundaries of innovation, we find ourselves looking to the natural world for inspiration. The field of biomimetics unlocks the secrets of mammalian limbs, translating their efficiency into groundbreaking applications.

Prosthetics and orthotics: There are prosthetic limbs designed with biomimetic precision, providing not just functionality but a harmonious blend of form and aesthetics. Orthotics, too, have transformed, drawing from the elegance and resilience of natural limbs to offer unprecedented support and mobility.

Robotics and biomechanics: The world of robotics embraces the lessons written in the sinews and joints of mammals, creating machines that replicate the agility of cheetahs, the precision of chameleons, and the adaptability of octopuses. Biomechanics takes center stage as engineers unravel the secrets of nature's locomotion to enhance the efficiency and versatility of machines.

Human-animal interaction: In a world where boundaries between technology and biology blur, human-animal interaction reaches new heights. Biomimetic limbs bridge the gap, fostering communication and understanding between species. From therapy animals with prosthetic limbs to robot-assisted rehabilitation for injured wildlife, the intersection of human and animal life is a testament to the symbiotic relationship between technology and nature.

Ethical Considerations and Inclusivity

Considering ethics in prosthetics: As we forge ahead in the realm of artificial limbs, ethical questions arise. From the sourcing of materials to the implications of enhancing natural abilities, we must tread carefully, ensuring that our technological strides do not compromise ethical principles.

Promoting accessibility: The benefits of biomimetic limb technologies should not be confined to a select few. These life-changing technologies must be accessible to individuals from all walks of life. This involves addressing economic, cultural, and geographical barriers to ensure that the marvels of biomimetic limbs reach those who need them most.

Empowering lives through biomimetic limbs: There are individuals whose lives have been forever changed, not merely by the functionality of biomimetic limbs, but by the restoration of dignity, independence, and the joy of rediscovering the wonders of movement. Those cases serve as a beacon, guiding us through the potential of biomimetic limbs to empower, inspire, and redefine the boundaries of what is possible.

Captivating Vision—Biomimetics of the Human Retina

The eyes, the window to the soul, have long captivated the imagination and curiosity of both poets and scientists alike. It's time for the retina to emerge as a masterpiece, a canvas woven with threads of remarkable functionality and breathtaking innovation. The captivating anatomy and function of the retina is a realm where photoreceptor cells dance to the rhythm of light and the symphony of vision orchestrates its grand performance.

Photoreceptor cells—The light symphony conductors: Picture the retina as a grand concert hall, where photoreceptor cells take center stage. Rods and cones, the virtuosos of the visual symphony, convert photons into electrical signals, setting the stage for the visual spectacle to unfold. Their sensitivity to different wavelengths transforms light into the language of the nervous system, a poetic dance that paints the world in vivid strokes.

Signal processing—The art of visual choreography: Beyond the initial notes played by photoreceptors, the retina choreographs a dazzling ballet of signal processing. Bipolar cells pass the baton, shaping the narrative of visual perception. Ganglion cells, the storytellers of this sensory saga, carry the refined signals to the brain, creating a narrative rich in detail and nuance. Witness the elegance of this visual choreography as the retina transforms raw data into a masterpiece of sight.

Color vision—A kaleidoscope of possibilities: The retina's ability to perceive color is a marvel in itself. Cones, the color connoisseurs, discern a spectrum that transcends ordinary perception. Humans have trichromatic vision, where reds, blues, and greens blend in a harmonious palette, creating a symphony of colors that enrich our visual experience.

Biomedical and Technological Insights—Healing Insights from Nature's Canvas

Mimicking nature's precision, scientists explore retinal regeneration and healing mechanisms, offering hope to those facing visual challenges. The retina is a source of inspiration for medical marvels and encourages us to push the boundaries of healing through biomimicry.

Advancements in Artificial Vision—The Rise of Technological Marvels

Artificial vision is a realm where human ingenuity mirrors the retina's brilliance. There are bionic eyes and retinal implants, technological wonders that illuminate the path toward restoring vision. Right now we can witness the blending of man and machine, where the retina's legacy fuels advancements that once seemed confined to the realms of science fiction.

Optical technologies—Crafting lenses from nature's blueprint: From camera lenses to telescopes, the retina's intricate architecture guides the creation of cutting-edge optics, pushing the boundaries of clarity and precision.

Innovations in imaging—Painting portraits with light: The advancement of imaging technologies is inspired by the retina's principles that shape the creation of vivid portraits and detailed scans.

The Marvel of Skin Regeneration

Our skin, the resilient cloak that shields us from the world, possesses an astonishing capacity for renewal, healing, and, perhaps most intriguingly, regeneration.

Skin is the body's protective canvas and is a testament to the resilience and capabilities inherent in mammalian life.

Cellular renewal—A dance of rejuvenation: At the heart of this captivating tale is the intricate ballet of cellular renewal. Like meticulous choreography, our skin constantly sheds old, worn-out cells, making room for the emergence of vibrant, new ones. This perpetual cycle of renewal not only maintains the skin's youthful glow but also serves as a testament to the body's extraordinary ability to rejuvenate itself.

Wound-healing mechanisms—Stitching the fabric of resilience: When wounds puncture our protective barrier, the skin enlists an impressive arsenal of healing mechanisms. From blood clotting to the orchestrated migration of fibroblasts, our body orchestrates a symphony of reparative processes. Witnessing the mending of a wound is like observing nature's magic show, where cellular actors take center stage to knit the torn fabric of our skin back together.

Scar reduction—Nature's vanishing act: In the grand narrative of regeneration, scars often linger as poignant reminders of past injuries. However, even here, the tale takes a fascinating turn. Nature, it seems, has its own vanishing act. Scar tissue, once considered a permanent mark, undergoes a dynamic metamorphosis, gradually fading away as the skin's regenerative ballet continues its performance.

Biomimetic Applications—Healing Beyond Nature's Reach

Inspired by the skin's innate ability to mend itself, researchers delve into the realm of biomimetics, exploring ways to harness this natural prowess for the benefit of materials and, intriguingly, even beyond the realm of biology. Imagine plastics equipped with a healing touch, bleeding chemicals that, when mixed, form a gel to mend damage. This biomimetic innovation not only mirrors the skin's regenerative magic but extends it into the realm of synthetic materials.

Biomedical applications—Tissue engineering's grand stage: The tale of skin regeneration unfolds on the grand stage of biomedical applications. Tissue engineering, propelled by our understanding of skin's regenerative secrets, emerges as a frontier where science and nature collaborate to heal wounds and even reconstruct entire skin layers. Advanced dressings become like transformative garments, aiding the body's natural regenerative processes.

Dermatology advancements—Nurturing the skin symphony: In the world of dermatology, advancements in skincare mirror the natural rhythms of the skin. Products designed to enhance cellular renewal, promote wound healing, and minimize scarring echo the harmonious melodies of nature's regenerative orchestra.

Ethical considerations—Navigating the moral compass: As we unravel the marvels of skin regeneration, ethical considerations beckon. In the pursuit of scientific progress, we must tread carefully, addressing the ethical implications of research and ensuring that innovations in dermatology adhere to a moral compass that prioritizes both human well-being and environmental sustainability.

A Symphony of Flavors— Biomimetics of Human Sense of Taste

In the grand orchestra of our sensory experience, the sense of taste takes center stage, orchestrating a delightful dance of flavors that transcends mere sustenance. Scientists and chefs alike draw inspiration from the diverse palates of mammals to craft an exquisite symphony of flavors.

Taste bud varieties—A palate of diversity: Within the confines of your mouth lies a universe of taste buds, each a specialized artist with a unique palette. From the sweet serenades to the savory ballads, taste buds distinguish themselves with an artistry that mirrors the diversity of the culinary world. Uncover the secrets of taste bud varieties, where bitter battles, sweet harmonies, and umami undertones converge in a gastronomic masterpiece.

The brain's role in flavor perception—The maestro of sensation: Behind every delectable sensation lies the maestro—your brain. The brain's role in flavor perception unveils the intricate ballet between taste buds and neurons, where the orchestra of senses orchestrates a performance that transcends the mere act of eating.

Taste involves more than your mouth—A multisensory extravaganza: Taste extends far beyond the boundaries of your mouth, intertwining with other senses in a multisensory extravaganza. Aroma, texture, and even the visual presentation of a dish join forces to create a sensory masterpiece. In the realm of taste, the experience is not merely confined to the palate but expands into a multisensory feast that captivates the mind and soul.

The art and science of artificial olfaction—A nose for innovation: Scientists have ventured into the creation of an electrochemical "nose," a marvel that mimics the olfactory wonders of the human sense of smell. Yet, any artificial device pales in comparison to the biological elegance that nature has sculpted.

Culinary and pharmaceutical implications—From kitchen to clinic: The exploration of taste transcends the kitchen, extending its reach into the realms of both culinary innovation and pharmaceutical marvels. Uncover the implications of taste in the culinary arts and pharmaceutical breakthroughs, where flavors become not only a source of delight but also a powerful tool in shaping the future of food and medicine.

Medicinal applications—Healing through flavorful alchemy: Taste, often overlooked as a mere indulgence, unveils its healing potential in medicinal applications. The alchemy of flavors intertwines with medicinal properties, offering not only a feast for the senses but also a prescription for well-being.

Enhancing food and beverage industries—A culinary revolution: As science unravels the secrets of taste, industries ranging from haute cuisine to mass production undergo a culinary revolution.

The limbs, vision, skin, and taste of mammals intertwine with human ingenuity, creating a harmonious blend of nature's wisdom and technological innovation. As we continue to uncover the secrets hidden within the biological wonders of mammals, curiosity knows no bounds, and the possibilities of biomimicry echo through the corridors of possibility.

Conclusion

Our journey has been a thrilling exploration of the incredible abilities that define nature, from the intricate noses of dogs and the majestic trunks of elephants to the echolocation prowess of bats and the brain-freezing tactics of Arctic squirrels.

Each intricate design nature has bestowed upon mammals—whether for survival, communication, or navigation—serves as a testament to the brilliance of the natural world. The trunks, noses, and sonar all play crucial roles in the survival strategies of these remarkable beings.

In contemplating the interplay between nature and technology, we recognize the inspiration that these incredible creatures provide for innovation. Biomimicry holds the potential to revolutionize our approach to problem-solving. From sonar technology inspired by bats to robotics inspired by the fine motor skills of an elephant's trunk, the synergy between nature and technology opens new frontiers of possibility.

Reflecting on these insights shows us that the interconnectedness of these abilities is vital for maintaining ecological balance. The delicate harmony between species, habitats, and ecosystems is a testament to the delicate dance orchestrated by nature. Mammals, in their myriad forms, contribute to this delicate balance, and their abilities are integral to the sustainability of the planet.

As we marvel at the wonders of mammals, we are compelled to become active participants in the conservation of our planet. The lessons learned from biomimicry extend beyond technological advancements; they encompass a deeper understanding of our interconnectedness with the environment and invite us to embrace a more profound connection with the natural world.

As we conclude this exploration, let's embrace a responsibility to appreciate and contribute to conservation efforts. The preservation of habitats, the protection of endangered species, and the mindful stewardship of our planet are essential tasks that require collective effort.

The significance of understanding and appreciating these incredible abilities extends beyond the pages of a book; it is a call to action. As stewards of this planet, we have the power to shape a future where our innovations coexist harmoniously with the wonders of the natural world. May our journey through the marvels of mammals inspire a collective commitment to preserve, protect, and cherish the interconnected web that sustains us all.

Let's remember that in the world of mammals, every ability is a piece of nature's grand design. Our understanding of these abilities is not only an intellectual pursuit but also a key to preserving the wonders that surround us. The legacy of these incredible creatures is in our hands, and it is through appreciation, understanding, and conservation that we can ensure the enduring marvels of mammals for generations to come.

References

Arctic ground squirrels changing hibernation patterns. (2023). USDA. https://www.fs.usda.gov/research/rmrs/news/releases/arctic-ground-squirrels-changing-hibernation-patterns

Arnold, C. (2018, November 19). *How cat tongues work—and can inspire human tech.* Animals. https://www.nationalgeographic.com/animals/article/understanding-cat-tongues-papillae

AskNature Team. (n.d.). *Skin resists microorganisms: Long-finned pilot whale.* AskNature. https://asknature.org/strategy/skin-resists-microorganisms

Averill, G. (2016, October 24). *Fur is the future of wetsuits.* Outside Online. https://www.outsideonline.com/outdoor-gear/water-sports-gear/your-next-wetsuit-will-mimic-otter-fur/

Bar-Cohen, Y. (2006). *Biomimetics – Using nature to inspire human innovation.* Research Gate. https://www.researchgate.net/publication/6168631_Biomimetics_-_Using_nature_to_inspire_human_innovation

Basit, A., et al. (2015). *Graceful gait transitions for biomimetic locomotion - The worm.* Research Gate. https://www.researchgate.net/publication/282 795692_Graceful_gait_transitions_for_biomime tic_locomotion_-_The_worm

Biomaterials for deep sea exploration engineering durable and sustainable solutions. (2023). Utilities One. https://utilitiesone.com/biomaterials-for-deep-sea-exploration-engineering-durable-and-sustainable-solutions

Biomimetics: Design inspired by human and animal bodies. (2015). Helen Cowan. https://www.helencowan.co.uk/biomimetics-design-inspired-human-and-animal-bodies

Biomimicry: Echolocation in robotics. (n.d.). Teach Engineering. https://www.teachengineering.org/activities/vi ew/nyu_biomimicry_activity1

Blasiak, et. al. (2022). *A forgotten element of the blue economy: Marine biomimetics and inspiration from the deep sea.* Research Gate. https://www.researchgate.net/publication/363 725148_A_forgotten_element_of_the_blue_eco nomy_Marine_biomimetics_and_inspiration_fr om_the_deep_sea

Bryce, E. (2023). *Marine "biomimetics" could be blue economy's next big hit.* China Dialogue Ocean. http://chinadialogueocean.net/en/conservation /marine-biomimetics-could-be-blue-economys-next-big-hit/https://www.epo.org/en/news-

events/press-centre/press-
release/2018/452056

Campo, R., et al.. (2021). Wine experiences: A review
from a multisensory perspective. *Applied Sciences*,
11(10), 4488.
https://doi.org/10.3390/app11104488

Castillote, L. (2024). *GEC 21 people and the earth's
ecosystem.* Course Hero.
https://www.coursehero.com/file/116185846/
GEC-21-Packet/

Dagenais, P., et a.. (2021). Elephants evolved strategies
reducing the biomechanical complexity of their
trunk. *Current Biology*, *31*(21), 4727-4737.e4.
https://doi.org/10.1016/j.cub.2021.08.029

Daley, J. (2018). Your cat's tongue is a rough, pink
engineering marvel. Smithsonian Magazine.
https://www.smithsonianmag.com/smart-
news/your-cats-tongue-masterpiece-design-
180970859/

Diebold, C. A., et al. (2020). Adaptive echolocation and
flight behaviors in bats can inspire technology
innovations for sonar tracking and interception.
Sensors, *20*(10), 2958.
https://doi.org/10.3390/s20102958

Drew, K. L., et al. (2016, February 26). Arctic ground
squirrel neuronal progenitor cells resist oxygen
and glucose deprivation-induced death. *World
Journal of Biological Chemistry*, *7*(1), 168.
https://doi.org/10.4331/wjbc.v7.i1.168

Dolphin-inspired compact sonar for enhanced underwater acoustic imaging. (2023, January 19). NUS News. https://news.nus.edu.sg/dolphin-inspired-compact-sonar

Fish, F. E., & Battle, J. M. (1995). Hydrodynamic design of the humpback whale flipper. *Journal of Morphology*, *225*(1), 51–60. https://doi.org/10.1002/jmor.1052250105

Fish, F. E., et al. (2011). The tubercles on humpback whales' flippers: Application of bio-inspired technology. *Integrative and Comparative Biology*, *51*(1), 203–213. https://doi.org/10.1093/icb/icr016

Fun facts about Super Sniffers. (2024). Menards. https://www.menards.com/main/pet-center/pet-center-archive/fun-facts-about-super-sniffers/c-7429806126845509.htm

How studying arctic ground squirrels can help advance human brain health. (2023, December 3). PBS NewsHour. https://www.pbs.org/newshour/show/how-studying-arctic-ground-squirrels-can-help-advance-human-brain-health

Jabr, F. (2012, June 26). *What the supercool Arctic ground squirrel teaches us about the brain's resilience.* Scientific American. https://www.scientificamerican.com/article/arctic-ground-squirrel-brain

Jean, G. (2008). *Dolphin's brain holds secret to more sophisticated sonar.* National Defense Magazine. https://www.nationaldefensemagazine.org/arti cles/2008/4/1/2008april-dolphins-brain-holds-secret-to-more-sophisticated-sonar

Kim, J.-J., et al. (2020). Biologically inspired artificial eyes and photonics. *Reports on Progress in Physics, 83*(4), 047101. https://doi.org/10.1088/1361-6633/ab6a42

Losey, K. (2019, September 11). *The future of innovation is here: 8 inventions from nature's laboratory.* Biomimicry Institute. https://biomimicry.org/the-future-of-innovation-is-here-8-inventions-from-natures-laboratory/

Love, D. (2014). *Your cat holds the secret to the future of robotics.* Business Insider. https://www.businessinsider.com/robot-e-whiskers-2014-1

Maderer, J. (2022). *Skin: An additional tool for the versatile elephant trunk.* Georgia Institute of Technology. https://www.gatech.edu/news/2022/07/18/sk in-additional-tool-versatile-elephant-trunk

McConville, E. (2019, February 7). *In biomimetic experiment, cat tongues lick the competition nearly every time.* Bates. https://www.bates.edu/news/2019/02/07/in-biomimetic-experiment-cat-tongues-lick-the-competition-nearly-every-time mrBrown. (2005, January 17). *Minimizing drag when flow over plate with hole*. Physics Forums. https://www.physicsforums.com/threads/mini mizing-drag-when-flow-over-plate-with-hole.60059/

Natarajan, K., et al. (2014). *Experimental studies on the effect of leading edge tubercles on laminar separation bubble*. Research Gate. https://www.researchgate.net/publication/269 248671_Experimental_Studies_On_The_Effect _Of_Leading_Edge_Tubercles_On_Laminar_S eparation_Bubble

Oyen, M. (2015, August 5). *Artificial whiskers could inspire future instruments to aid keyhole surgery*. The Conversation. https://theconversation.com/artificial-whiskers-could-inspire-future-instruments-to-aid-keyhole-surgery-45670

Pavid, K. (n.d.). *Secrets of the deepest diving whales*. Natural History Museum. https://www.nhm.ac.uk/discover/secrets-of-deep-diving-whales.html

Pena, B., & Muk-Pavic, E. (2019). *Biomimetics in ship design?* International Conference on Ships and Offshore Structures. https://discovery.ucl.ac.uk/id/eprint/10086107 /7/Muk-Pavic_Biomimetics%20in%20Ship%20Design_ AAM.pdf petdog347. (2020, June 15). *The skin of the long-finned pilot whale.* Reddit. https://www.reddit.com/r/Awwducational/co mments/h96z8o/the_skin_of_the_longfinned_ pilot_whale_has/

Regal, S., et al. (2021). *Biomimetic models of the human eye, and their applications.* Research Gate. https://www.researchgate.net/publication/350 555748_Biomimetic_models_of_the_human_ey e_and_their_applications

Riordan, K., et al. (2023). Ontogenetic changes in southern sea otter (*Enhydra lutris nereis*) fur morphology. *Journal of Morphology, 284*(9). https://doi.org/10.1002/jmor.21624

Sea otter fur. (n.d.). Listen in English. https://www.listeninenglish.com/av020-sea-otter-fur.php

Spence, C. (2020). Multisensory flavour perception: Blending, mixing, fusion, and pairing within and between the senses. *Foods, 9*(4), 407. https://doi.org/10.3390/foods9040407

Steroids: the arctic ground squirrel's secret to surviving winter. (2014). Canadian Geographic. https://canadiangeographic.ca/articles/steroids -the-arctic-ground-squirrels-secret-to-surviving-winter

Sun, Q., et al. (2022). Microstructure and self-healing capability of artificial skin composites using biomimetic fibers containing a healing agent.

Polymers, *15*(1), 190. https://doi.org/10.3390/polym15010190

Super Sniffers – Live! (n.d.). Medical Detection Dogs. https://www.medicaldetectiondogs.org.uk/super-sniffers-live

Super-Sniffers (n.d.). Earth Date. https://www.earthdate.org/episodes/super-sniffers

The potential of biomimicry in wind turbine blade aerodynamics. (2023). Utilities One. https://utilitiesone.com/the-potential-of-biomimicry-in-wind-turbine-blade-aerodynamics

Turbines and fans inspired by whales: Stephen Dewar, Philip Watts and Frank Fish named European Inventor Award 2018 finalists. (2018, April 24). European Patent Office https://www.epo.org/en/news-events/press-centre/press-release/2018/452056

Villa, E. (2024). *The Cuvier's beaked whale is the champion of deep diving.* CW Azores. https://www.cwazores.com/blog/Whales%20and%20Dolphins/ziphius-champion-of-deep-diving

Vincelette, A. (2023). The characteristics, distribution, function, and origin of alternative lateral horse gaits. *Animals*, *13*(16), 2557. https://doi.org/10.3390/ani13162557

Wang, B., & Sullivan, T. (2017). *A review of terrestrial, aerial and aquatic keratins: The structure and mechanical properties of pangolin scales, feather shafts and baleen plates.* PubMed. https://pubmed.ncbi.nlm.nih.gov/28522235

Wang, B., et al. (2016). *Pangolin armor: Overlapping, structure, and mechanical properties of the keratinous scales.* Research Gate. https://www.researchgate.net/publication/303 499715_Pangolin_armor_Overlapping_structur e_and_mechanical_properties_of_the_keratino us_scales

Wang, Z., et al. (2023). Bioinspired skin towards next-generation rehabilitation medicine. *Frontiers in Bioengineering and Biotechnology*, *11*, 1196174. https://doi.org/10.3389/fbioe.2023.1196174

Wu, C., et al. (2017). Biomimetic sensors for the senses: Towards better understanding of taste and odor sensation. *Sensors (Basel, Switzerland)*, *17*(12). https://doi.org/10.3390/s17122881

Yovel, Y., et al. (2008). Plant classification from bat-like echolocation signals. *PLoS Computational Biology*, *4*(3), e1000032. https://doi.org/10.1371/journal.pcbi.1000032

www.ingramcontent.com/pod-product-compliance
Lightning Source LLC
Chambersburg PA
CBHW032054040426
42335CB00037B/710